*Polymer Surfaces
and Interfaces II*

Polymer Surfaces and Interfaces II

Edited by

W. J. Feast

IRC in Polymer Science and Technology
University of Durham
Durham, UK

H. S. Munro

Courtaulds Research
Coventry, UK

and

R. W. Richards

IRC in Polymer Science and Technology
University of Durham
Durham, UK

JOHN WILEY & SONS

Chichester · New York · Brisbane · Toronto · Singapore

Other Wiley Editorial Offices

John Wiley & Sons, Inc., 605 Third Avenue,
New York, NY 10158-0012, USA

Jacaranda Wiley Ltd, G.P.O. Box 859, Brisbane,
Queensland 4001, Australia

John Wiley & Sons (Canada) Ltd, 22 Worcester Road,
Rexdale, Ontario M9W 1L1, Canada

John Wiley & Sons (SEA) Pte Ltd, 37 Jalan Pemimpin #05-04,
Block B, Union Industrial Building, Singapore 2057

Library of Congress Cataloging-in-Publication Data

Polymer Surfaces and Interfaces II / edited by W. J. Feast, H. S. Munro,
and R. W. Richards.
    p.   cm.
    Includes bibliographical references and index.
    ISBN 0 471 93456 9
    1. Polymers—Surfaces.   2. Surface Chemistry.   I. Feast, W. J.
II. Munro, H. S.   III. Richards, R. W. (Randal William), 1948–
QD381.9.S97P649   1992
620. 1'920429—dc20   92-30939
                                                            CIP

British Library Cataloguing in Publication Data

A catalogue record for this book is available from the British Library

ISBN 0 471 93456 9

Typeset by Techset Composition Ltd., Salisbury
Printed and bound in Great Britain by Biddles Ltd, Guildford, Surrey

Contents

3   Non-equilibrium Effects in Polymeric Stabilization   49

M. E. Cates and J. T. Brooks

**4   Ion Beam Analysis of Composition Profiles near Polymer Surfaces
and Interfaces**   71

R. A. L. Jones

7 **Surface Modification and Analysis of Ultra-high-modulus Polyethylene Fibres for Composites** 161

G. A. George

Contributors

ALLARA, D. L.   *Department of Materials Science and Department of Chemistry, Pennsylvania State University, University Park, Pennsylvania 16802, USA*

ATRE, S. V.   *Department of Materials Science and Department of Chemistry, Pennsylvania State University, University Park, Pennsylvania 16802, USA*

BEE, T. G.   *Polymer Science and Engineering Department, University of Massachusetts, Amherst, Massachusetts 01003, USA*

BROOKS, J. T.   *Theory of Condensed Matter, Cavendish Laboratory, Madingley Road, Cambridge, CB3 0HE, UK*

CATES, M. E.   *Theory of Condensed Matter, Cavendish Laboratory, Madingley Road, Cambridge, CB3 0HE, UK*

DAVIES, M. C.   *Department of Pharmaceutical Sciences, University of Nottingham, Nottingham, NG7 2RD, UK*

DIAS, A. J.   *Polymer Science and Engineering Department, University of Massachusetts, Amherst, Massachusetts 01003, USA*

EARNSHAW, J. C.   *Department of Pure and Applied Physics, The Queen's University of Belfast, Belfast, BT7 1NN, UK*

FRANCHINA, N. L.   *Polymer Science and Engineering Department, University of Massachusetts, Amherst, Massachusetts 011003, USA*

GEORGE, G. A.   *Department of Chemistry, University of Queensland, Queensland 4072, Australia*

IMANISHI, Y.   *Department of Polymer Chemistry, Kyoto University, Yoshida Honmachi, Sakyo-ku, Kyoto, 606 Japan*

ITO, Y.   *Department of Polymer Chemistry, Kyoto University, Yoshida Honmachi, Sakyo-ku, Kyoto, 606 Japan*

JACKSON, D. E.   *The VG STM Laboratory for Biological Applications, Department of Pharmaceutical Sciences, University of Nottingham, University Park, Nottingham, NG7 2RD, UK*

JONES, R. A. L.   *PCS, Cavendish Laboratory, Madingley Road, Cambridge, CB3 0HE, UK*

KOLB, B. U.   *Polymer Science and Engineering Department, University of Massachusetts, Amherst, Massachusetts 01003, USA*

KREUSEL, K. M.   *The VG STM Laboratory for Biological Applications,*
                     Department of Pharmaceutical Sciences, University of
                     Nottingham, University Park, Nottingham, NG7 2RD, UK

LEE, K.-W.   *Polymer Science and Engineering Department, University*
                     of Massachusetts, Amherst, Massachusetts 01003, USA

LIU, L.-S.   *Department of Polymer Chemistry, Kyoto University,*
                     Yoshida Honmachi, Sakyo-ku, Kyoto, 606 Japan

McCARTHY, T. J.   *Polymer Science and Engineering Department, University*
                     of Massachusetts, Amherst, Massachusetts 01003, USA

PARIKH, A. N.   *Department of Materials Science and Department of*
                     Chemistry, Pennsylvania State University, University Park,
                     Pennsylvania 16802, USA

PATTON, P. A.   *Polymer Science and Engineering Department, University*
                     of Massachusetts, Amherst, Massachusetts 01003, USA

ROBERTS, C. J.   *The VG STM Laboratory for Biological Applications,*
                     Department of Pharmaceutical Sciences, University of
                     Nottingham, University Park, Nottingham, NG7 2RD, UK

SHOICHET, M. S.   *Polymer Science and Engineering Department, University*
                     of Massachusetts, Amherst, Massachusetts 01003, USA

TENDLER, S. J. B.   *The VG STM Laboratory for Biological Applications,*
                     Department of Pharmaceutical Sciences, University of
                     Nottingham, University Park, Nottingham, NG7 2RD, UK

VAN OSS, C. J.   *Departments of Microbiology and Chemical Engineering,*
                     State University of New York at Buffalo, Buffalo, New
                     York 14214, USA

WILKINS, M. J.   *The VG STM Laboratory for Biological Applications,*
                     Department of Pharmaceutical Sciences, University of
                     Nottingham, University Park, Nottingham, NG7 2RD, UK

WILLIAMS, P. M.   *The VG STM Laboratory for Biological Applications,*
                     Department of Pharmaceutical Sciences, University of
                     Nottingham, University Park, Nottingham, NG7 2RD, UK

YOUNG, R. J.   *Polymer Science and Technology Group, Manchester*
                     Materials Science Centre, UMIST/University of Man-
                     chester, Manchester, M2 7HS, UK

Preface

This book is the third in an irregular series on the theme of polymer surfaces and interfaces. Like the earlier books, *Polymer Surfaces* edited by D. T. Clark and W. J. Feast (Wiley, 1978) and *Polymer Surfaces and Interfaces* edited by W. J. Feast and H. S. Munro (Wiley, 1987), this volume is a compilation of chapters written by contributors to an international symposium held in Durham. The symposium took place during July 1991 under the auspices of the Pure and Applied Macromolecular Chemistry Group of the Royal Society of Chemistry and the Society of Chemical Industry. Durham University acted as hosts for the meeting and financial sponsorship was provided by BP, British Gas, Cookson Group, Courtaulds, European Research Office (US Army), Hydro Polymers, ICI, Pilkingtons, Unilever and VG Instruments.

These symposia have been based on a single sequence of lectures presented by acknowledged experts in their fields and given an audience drawn from a wide range of scientific disciplines, and with a variety of perspectives concerning the importance of polymer surfaces and interfaces. Speakers were allowed time to develop their subjects and encouraged to bear in mind that the audience was, for the most part, not expert in their speciality. This format leads to many questions and a lively debate and these meetings have a good track record for initiating a new interdisciplinary research collaborations.

In the intervening time since publication of the 1987 volume, new methods of interrogating polymer surfaces have been developed. Two of these, ion beam analysis and surface quasi-elastic light scattering are dealt with here. Additionally, theories of polymers attached to solid surfaces have been vigorously pursued and the survey of these (by Cates) provides a link to the experimental work discussed by Luckham in the 1987 volume. All the speakers were invited to prepare chapters based on their presentations; just over half accepted the invitation and this book is the result of their labours. Some speakers were unable to prepare manuscripts because of the pressure of other commitments and some felt that their views had already received sufficient exposure in the recent past. The objective and style of the meeting is reflected in the chapters presented here and since the chapters are mainly of an overview kind rather than a report of current work we hope that the delay between the meeting and the publication of this book will not detract from its value to the reader. The previous volumes in this series have been well received and we hope that this volume will make its contribution to advancing knowledge and understanding in this important field.

W. J. Feast
H. S. Munro
R. W. Richards

1

Surface Chemistry of Chemically Resistant Polymers

Timothy G. Bee, Anthony J. Dias, Nicole L. Franchina, Brant U. Kolb, Kang-Wook Lee, Penelope A. Patton, Molly S. Shoichet and Thomas J. McCarthy

Polymer Science and Engineering Department
University of Massachusetts

1 INTRODUCTION

We have been involved in a research program with the long-range objective of polymer surface property control (e.g. wettability, adhesion, coefficient of friction). We wish to be able to both predict surface properties with the knowledge of surface structure and control surface properties by manipulating structure using organic surface chemistry. As a first step towards this objective, we have developed methods to introduce reactive organic functional groups on to the surfaces of unreactive polymer film substrates. This chapter reviews portions of our work on poly(chlorotrifluoroethylene) (PCTFE), poly(tetra-fluoroethylene-co-hexafluoropropylene) (FEP), poly(vinylidene fluoride) (PVF$_2$) and poly(ether ether ketone) (PEEK). The fluoropolymer films function as inert supports for their reactive functionality once the film surface is modified. Virgin semicrystalline PEEK contains a reactive functional group, the diaryl ketone. We discuss here methods for introducing reactive functionality to these fluoro-polymers as well as selected transformations of both the fluoropolymer surface functional groups and PEEK diaryl ketones.

To correlate surface structure with surface properties we need a series of polymer film samples that differ only in the identity of their surface functional groups. Our strategy is to prepare, in an initial modification step, a substrate (pictured generically in Figure 1) with a thin ('submonolayer' level to several 'monolayer' level) layer of versatile reactive groups. These groups (X) can then be converted by chemical reactions to yield the series of film samples to be studied. The reason for the choice of chemically resistant polymers is obvious: we want to be able to use a range of reaction conditions (some quite brutal) and not further affect the underlying substrate. A problem that must be avoided

Polymer Surfaces and Interfaces II
Edited by W. J. Feast, H. S. Munro and R. W. Richards
© 1993 John Wiley & Sons Ltd

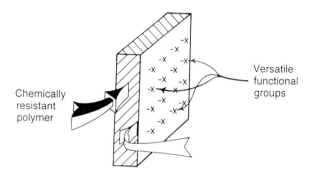

Figure 1. Target modified polymer film substrates (upper). (Reprinted with permission from *Macromolecules*, **20**, 2068. Copyright (1987) American Chemical Society)

in the initial surface modification step, which is inherent to reactions of chemically resistant polymers at interfaces with reactive solutions, is their tendency to react by a corrosive reaction. If the product is more reactive or if it is swollen by the reaction solution to a greater extent than the virgin polymer, a pitted surface that is not amenable to structure–property correlations (Figure 2) will result.

To prepare a modified polymer surface of the type in Figure 1 by reaction of a solid polymer at a solid–solution interface and avoid the scenario of Figure 2, a number of factors should be considered, particularly in regard to controlling the surface selectivity of the reaction. The virgin solid polymer in contact with

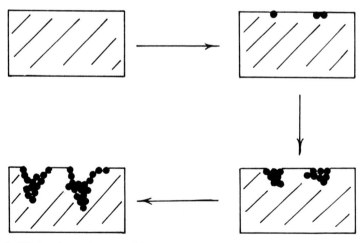

Figure 2. Pitting in surface modification reactions of chemically resistant polymers. (Reprinted with permission from Dias and McCarthy, *Macromolecules*, **17**, 2529. Copyright (1984) American Chemical Society)

the reactive solution will interact with the solvent and reagent(s) to varying extents ranging from not being wet to being highly swollen. The reagent(s) may partition between the condensed phases to varying extents. The interface will thus vary from sharp to diffuse and therefore the thickness and homogeneity (a gradient is possible) of the modified layer in the product will be affected by the nature of this interface. The reaction temperature can also affect the diffuseness of the interface. Upon reaction a different polymer–solution interface results and the conditions for the reaction change. The partially modified polymer surface can interact with the solution to a greater or lesser extent than does the unreacted polymer surface. This can lead to processes ranging from corrosion (an unreacted polymer which is barely wet by the solvent converting to a modified polymer which is soluble) to autoinhibition (the product acts as a barrier layer and protects the underlying material from reaction). The oxidation of polypropylene[1] with CrO_3 in acetic acid/acetic anhydride is an example of the former; the dehydrofluorination of PVF_2 with $Bu_4N^+OH^-$ in aqueous $NaOH$[2] is an example of the latter. Additionally, the specific chemistry that occurs can affect the physical structure of the product. If the solution interacts with the product to a great extent, but the modification chemistry involves crosslinking, the product will not dissolve and a thick modified layer will result. If the modification chemistry involves polymer chain cleavage, low molecular weight material will dissolve and chain ends will be important chemical features. The oxidation of polyethylene with chromic acid[3,4] to form a surface rich in carboxylic acids is an example of this. A further matter for consideration is that of the differences between the structures of the product in contact with the reaction solution and when isolated from the reaction solution. The compatibility (or lack thereof) of the unreacted and modified polymer and the differences between surface and interfacial free energies may induce surface reconstructions[5] during rinsing procedures and solvent removal.

2 POLY(CHLOROTRIFLUOROETHYLENE) (PCTFE)

PCTFE film reacts with alkyllithium reagents at the solid polymer–solution interface to incorporate the alkyl groups of the lithium reagent on polymer chains near the surface of the film:

(1)

The mechanism was studied using PCTFE film, oil and high surface area powder.[6] The first step is a metal–halogen exchange reaction to form a chloroalkane and an unstable lithiated polymer. Lithium fluoride is eliminated to yield a difluoroolefin. Then a second equivalent of alkyllithium adds to the difluoroolefin, lithium fluoride is eliminated and the alkyl-substituted olefin is formed.

The following equations describe the strategy used to introduce carboxylic acids, alcohols and aldehydes to the PCTFE surface:[6]

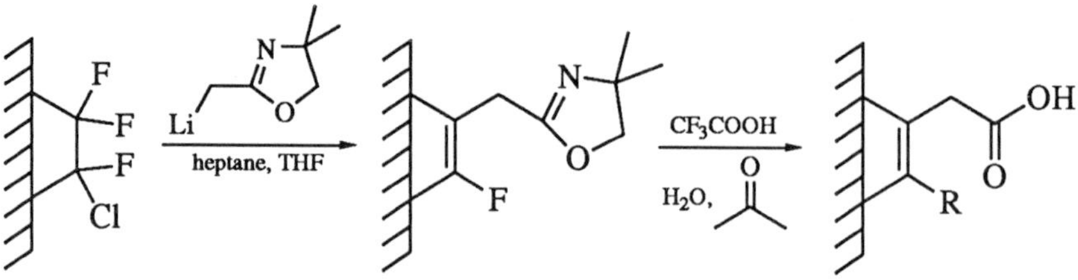

(2)

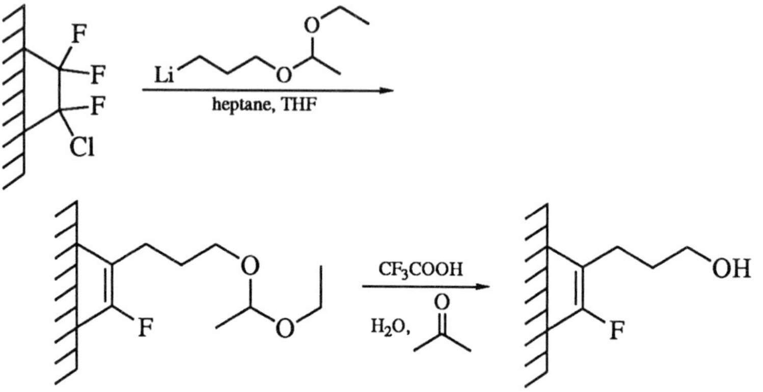

(3)

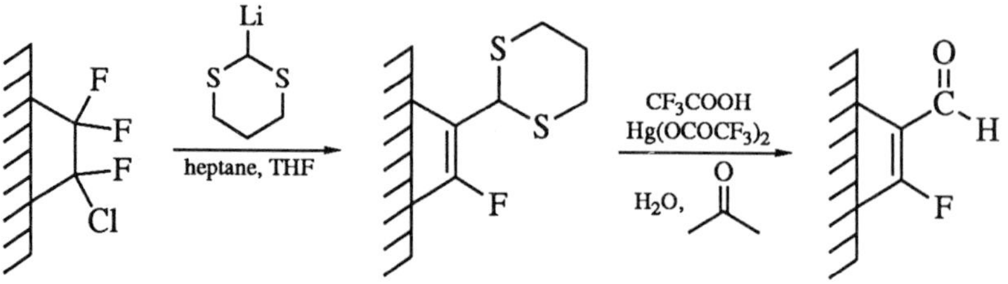

(4)

2-Lithiomethyl-4,4-dimethyloxazoline ($Li[CH_2CO_2H]^P$) contains a protected carboxylic acid, acetaldehyde 3-lithiopropyl ethyl acetal ($Li[(CH_2)_3OH]^P$) contains a protected alcohol and 2-lithiodithiane ($Li[CHO]^P$) contains a protected aldehyde.

The kinetics for the reaction of PCTFE film with $Li[(CH)_2CO_2H]^P$ in 50:50 THF (tetrahydrofuran)/heptane at three temperatures were obtained using

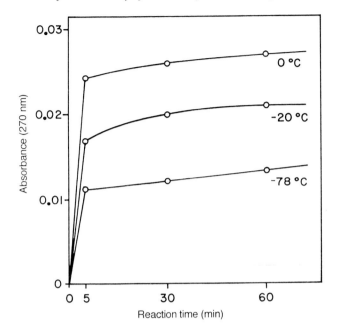

Figure 3. Plots of absorbance (270 mm) versus reaction time for the reactions of PCTFE film with $Li[CO_2H]^P$ in 50:50 heptane/THF at 0, -20 and $-78°C$. (Reprinted with permission from *Macromolecules*, **20**, 2068. Copyright (1987) American Chemical Society)

UV-vis (ultraviolet–visible) spectroscopy (the absorbance at 270 nm is due to unsaturation in the polymer chain) and the results are shown in Figure 3. The reactions are autoinhibiting and essentially complete after 5 minutes. This indicates that the product film surface does not interact extensively with the reaction solution and that the modified surface inhibits further reaction: the reaction is surface elective. X-ray photoelectron spectroscopy (XPS) of PCTFE–$[CH_2CO_2H]^P$ shows the expected nitrogen (402 eV) and oxygen (532 eV) photoelectron lines and decreased chlorine (208, 278 eV) and fluorine (685 eV) intensities. The carbon 1s region of the XPS spectrum of PCTFE–$[CH_2CO_2H]^P$ prepared at $-78°C$ for 60 minutes is shown in Figure 4. The upper spectrum was obtained at a 15° takeoff angle (measured between the plane of the film and the detector) and the lower spectrum at a 75° takeoff angle. These spectra indicate the composition of the outer ~ 10 and ~ 40 Å of the sample, respectively, and along with the kinetics suggest a very surface-selective reaction. The majority of the protected carboxylic acid functionality present is in the outer 10 Å. When the reaction PCTFE with $Li[CH_2CO_2H]^P$ is carried out at $-20°C$ for 60 minutes, the outer 10 Å contains no unreacted PCTFE, but a small amount remains in the outer 40 Å. At 0°C, the outer 40 Å is completely

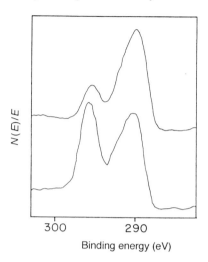

300   290

Binding energy (eV)

Figure 4. XPS spectra (C 1s region) for PCTFE–[CO$_2$H]^P prepared at $-78°C$ in 50:50 heptane/THF (reaction time 60 minutes). The takeoff angles used to obtain the spectra were 15° for the upper spectrum and 75° for the lower spectrum. (Reprinted with permission from *Macromolecules*, **20**, 2068. Copyright (1987) American Chemical Society)

reacted after 6 minutes. The reactions of PCTFE with Li[CHO]^P exhibit similar autoinhibitive kinetics and temperature dependence; the reaction depths are about twice as great as those for Li[CH$_2$CO$_2$H]^P under identical conditions.

The depth of each of these reactions (thickness of the modified layer) can be controlled conveniently with temperature. This temperature dependence is probably due to the greater mobility of surface chains at higher temperatures. With greater mobility, more reaction sites are exposed to reagents in solution and thicker modified layers result. This mobility also depends on the composition of the reaction solvent in contact with the polymer; THF wets and swells PCTFE to a greater extent than does heptane and higher THF content increases chain mobility and causes thicker modified layers. This effect was examined quantitatively for the reactions of Li[CHO]^P and Li[CH$_2$CO$_2$H]^P with PCTFE at $-20°C$ using different composition heptane/THF solvent. The UV results are summarized in Figure 5. The effect of solvent composition on reaction depth is more pronounced with Li[CHO]^P than with Li[CH$_2$CO$_2$H]^P and this indicates that solvent interaction with the product surface is the most important effect.

The kinetics for the reaction of PCTFE with Li[(CH$_2$)$_3$OH]^P were markedly different from those for Li[CHO]^P and Li[CH$_2$CO$_2$H]^P. The results shown in Figure 6 indicate that much deeper reactions and no autoinhibition occur. We estimate, using a combination of XPS and UV spectroscopies, that the modified

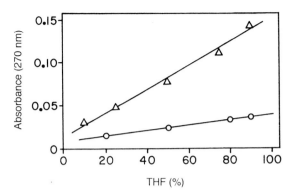

Figure 5. Plots of absorbance (270 nm) versus heptane/THF composition for the reactions of Li[CHO]^P and Li[CO$_2$H]^P with PCTFE film at $-20°$C for 60 minutes. (Reprinted with permission from *Macromolecules*, **20**, 2068. Copyright (1987) American Chemical Society)

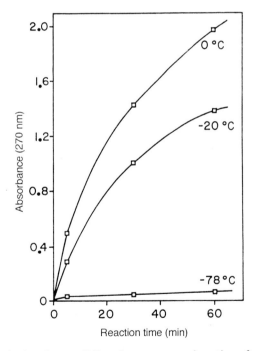

Figure 6. Plots of absorbance (260 nm) versus reaction time for the reaction of PCTFE with Li[(CH$_2$)$_3$OH]^P in 50:50 THF/heptane at -78, -20 and 0°C. (Reprinted with permission from *Macromolecules*, **20**, 2068. Copyright (1987) American Chemical Society)

layer thicknesses after 60 minutes reaction are 50, 1150 and 1600 Å at -78, -20 and $0°C$, respectively. These relatively deep reactions are most certainly the result of strong interactions between the reaction solution and the acetal modified surface.

The deprotections of the functional groups to give PCTFE–CH$_2$CO$_2$H, PCTFE–CHO and PCTFE–(CH$_2$)$_3$OH were not straightforward and a number of conditions were tried and optimized. Each deprotection involves hydrolysis and there is difficulty getting water to the protected functional groups on these relatively hydrophobic surfaces. The use of organic acid catalysis and acetone facilitated the hydrolysis reactions.

The reactivities of PCTFE–CH$_2$CO$_2$H and PCTFE–(CH$_2$)$_3$OH were assessed in detail using standard solution conditions for acid and alcohol transformations. The alcohol surface proved to be reactive and versatile (see below), but PCTFE–CH$_2$CO$_2$H turned out to be almost completely unreactive. A consistent explanation for the low reactivity is that the carboxylic acid is separated from the polymer backbone by only a single methylene unit. We have recently[7] prepared a reactive, densely functionalized carboxylic acid surface (PCTFE–(CH$_2$)$_3$CO$_2$H):

$$(5)$$

This should prove to be a complementary surface to PCTFE–(CH$_2$)$_3$OH for surface-chemical studies.

Living anionic polymers can be grafted to the PCTFE surface using this chemistry (equation 1). Butadiene–endcapped polystyrllithium in benzene reacts with PCTFE to form a thin grafted overlayer of polystyrene on PCTFE.[8] Figure 7 shows XPS and attenuated total reflectance infrared (ATR IR) spectra of a grafted sample. The thickness of the polystyrene layer can be controlled with solvent composition, temperature, reaction time and polymer molecular weight.

Water contact angles (dynamic advancing, θ_A, and receding, θ_R) have been measured for each of the surfaces described above; the data are summarized in Table 1.

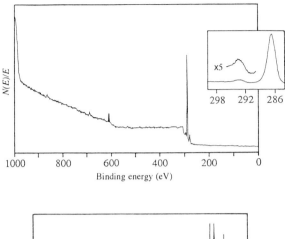

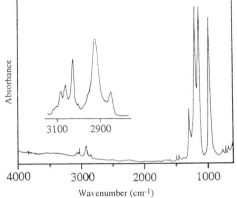

Figure 7. XPS and ATR IR spectrum of PCTFE film which had been reacted with butadiene–endcapped polystyryllithium (M_n = 5000, 0.024 M) in refluxing benzene. (Reprinted with permission from Kolb *et al.*, *Macromolecules*, **23**, 366. Copyright (1991) American Chemical Society)

PCTFE–$(CH_2)_3OH$ undergoes a wide range of alcohol transformations in essentially 100% yield.[9] We review here only esterification and urethanation reactions. One advantage of PCTFE as a substrate for organic surface chemistry is that it is amenable to standard surface analytical techniques; XPS and ATR IR are particularly powerful. Figures 8 and 9 exhibit survey XPS and ATR IR spectra of PCTFE, PCTFE–$[(CH_2)_3OH]^P$ and PCTFE–$(CH_2)_3OH$. The ATR IR spectrum of PCTFE has no absorbance in most regions of interest. Upon reaction with Li$[(CH_2)_3OH]^P$, CH_3 and CH_2 stretching (2978, 2936, 2899, 2876 cm^{-1}) and bending (1480, 1445, 1381 cm^{-1}) and C—O stretching (1063 cm^{-1}) modes are introduced. Upon hydrolysis to PCTFE–$(CH_2)_3OH$, the methyl bands disappear and a broad O—H stretching band (3335 cm^{-1})

Table 1. Water contact angle data for functionalized PCTFE

Sample	Contact angles (deg)	
	θ_A	θ_R
PCTFE	104	77
PCTFE–[CH$_2$CO$_2$H]^P	88	43
PCTFE–CH$_2$CO$_2$H	71	0
PCTFE–[CHO]^P	88	61
PCTFE–CHO	75	25
PCTFE–[(CH)$_3$OH]^P	73	42
PCTFE–(CH)$_3$OH	69	19
PCTFE–polystyrene	94	68
PCTFE–[(CH)$_3$CO$_2$H]^P	76	49
PCTFE–(CH)$_3$CO$_2$H	55	0

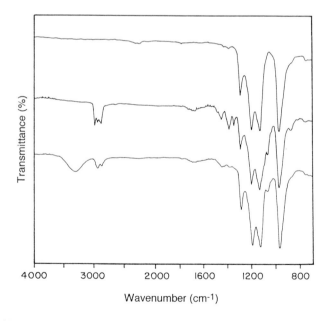

Figure 8. ATR IR and C 1s region spectra of PCTFE (top), PCTFE–[(CH$_2$)$_3$OH]^P (middle) and PCTFE–(CH$_2$)$_3$OH (bottom). (Reprinted with permission from Lea and McCarthy, *Macromolecules*, **21**, 2318. Copyright (1988) American Chemical Society)

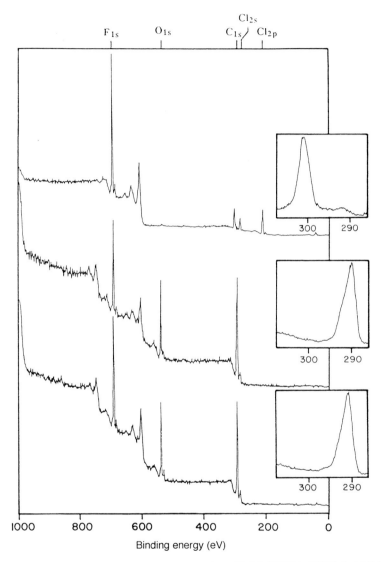

Figure 9. XPS survey spectra of PCTFE (top), PCTFE–[(CH$_2$)$_3$OH]^P (middle) and PCTFE–(CH$_2$)$_3$OH (bottom). (Reprinted with permission from Lea and McCarthy, *Macromolecules*, **21**, 2318. Copyright (1988) American Chemical Society)

is displayed. The XPS spectra are equally lucid. Upon reaction with
Li[(CH$_2$)$_3$OH]^P, the incorporation of oxygen, the partial removal of fluorine
and complete removal of chlorine are observed. The high binding energy C 1s
photoelectron line (301 eV—not corrected for charging) due to CF$_2$ and CFCl
in PCTFE is completely removed and replaced by the complex line at lower
binding energy (290 eV) due to all of the carbons in PCTFE–[(CH$_2$)$_3$OH]^P.
Quantitative XPS for this system proved quite useful. XPS atomic ratios for
PCTFE–[(CH$_2$)$_3$OH]^P were consistently 73:12:15 (C:F:O). The predicted
ratios are 75:8:17; thus PCTFE–[(CH$_2$)$_3$OH]^P exhibits lower values than
predicted for carbon and oxygen and a higher value than predicted for fluorine
based on the structure in equation (3). In that these differences are counter to
what would be expected from sample contamination or oxidation, we propose

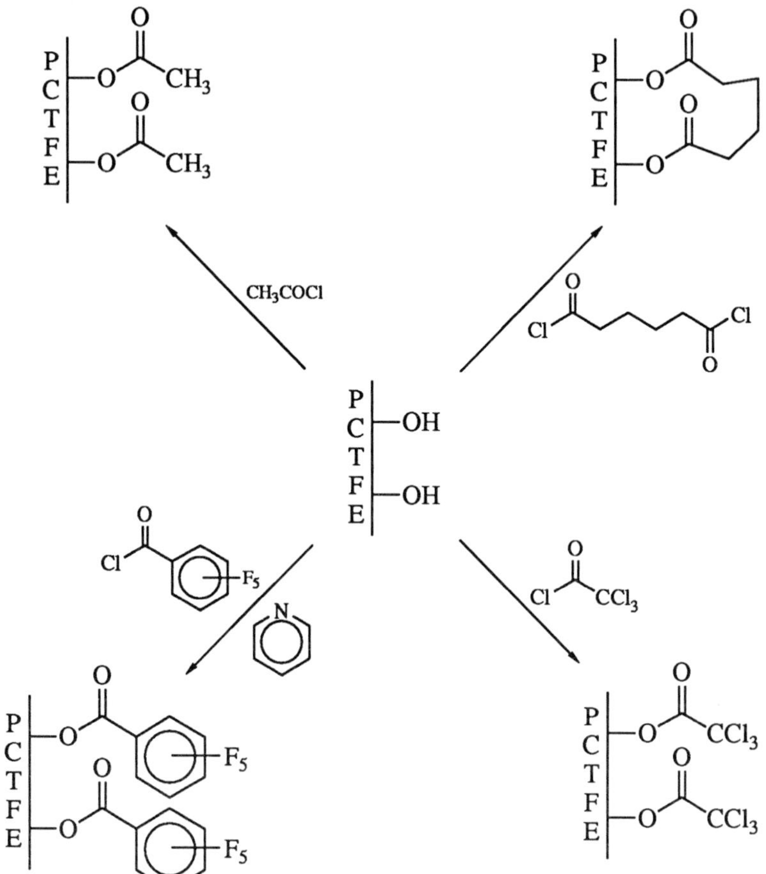

Scheme I

that PCTFE–[(CH$_2$)$_3$OH]^P consists of 80% modified repeat units and 20% difluoroolefins (the reaction intermediate). This structure has the observed atomic ratios (four protected functional groups per five polymer repeat units). The validity of this proposal is supported by much of the data for subsequent reactions.

Esterification reactions of PCTFE–(CH$_2$)$_3$OH with acetyl chloride, trichloro-acetyl chloride and adipoyl chloride (Scheme I) in THF at room temperature proceeded to completion in 12–40 hours. The reaction with pentafluorobenzoyl chloride was slower but was complete after 24 hours with the addition of pyridine. ATR IR spectra (Figure 10) indicate essentially complete reaction and the expected ester carbonyl bands for the acetate (1740 cm^{-1}), trichloroacetate (1767 cm^{-1}), adipate (1734 cm^{-1}) and pentafluorobenzoate (1742 cm^{-1}). The C 1s regions of the XPS spectra of the acetate, trichloroacetate and adipate could be nicely curve-fitted with three peaks. The C 1s region of the pentafluoro-benzoate was less well resolved and not suited for this treatment. Figure 11 shows these data along with a two-peak curve-fitted spectrum of PCTFE–(CH$_2$)$_3$OH for comparison. In each case the observed areas agree well with the values predicted for the proposed structures.

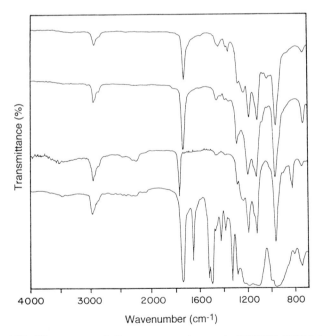

Figure 10. ATR IR spectra of (from top to bottom): PCTFE–(CH$_2$)$_3$OC(O)CH$_3$, PCTFE – [(CH$_2$)$_3$OC(O)CH$_2$CH$_2$]$_2$, PCTFE – (CH$_2$)$_3$OC(O)CCl$_3$ and PCTFE – OC(O)C$_6$F$_5$. (Reprinted with permission from Lea and McCarthy, *Macromolecules*, **21**, 2318. Copyright (1988) American Chemical Society)

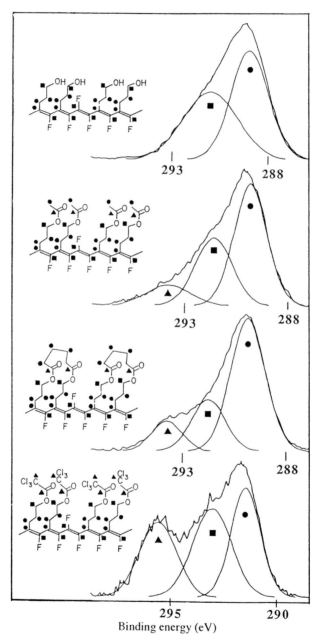

Figure 11. Carbon 1s regions of XPS spectra of (from top to bottom): PCTFE–(CH$_2$)$_3$OH, PCTFE – (CH$_2$)$_3$OC(O)CH$_3$, PCTFE – [(CH$_2$)$_3$OC(O)CH$_2$CH$_2$]$_2$ and PCTFE–(CH$_2$)$_3$OC(O)CCl$_3$. (Reprinted with permission from Lea and McCarthy, *Macromolecules*, **21**, 2318. Copyright (1988) American Chemical Society)

The reactions of PCTFE–(CH$_2$)$_3$OH with α,α,α-trifluoro-*p*-tolyl isocyanate and trichloroacetyl isocyanate in THF at room temperature (Scheme II) are complete after 12 hours to yield PCTFE–(CH$_2$)$_3$OC(O)NHC$_6$H$_4$-*p*-CF$_3$ and PCTFE–(CH$_2$)$_3$OC(O)NHC(O)CCl$_3$. The reaction with hexamethylene diisocyanate is considerably slower and takes 24 hours in THF in the presence of catalytic amounts of dibutyltin dilaurate to yield PCTFE–[OC(O)NHCH$_2$CH$_2$CH$_2$]$_2$. The contact angle data for these products and the ester surfaces are displayed in Table 2. The hydrophobicity of the ester and urethane surfaces varies in the order: fluorine-containing surface > chlorine-containing surface > non-halogenated surface.

ATR IR spectra (Figure 12) are consistent with urethane formation. This technique does not ascertain that the reactions are complete since the N—H bands appear in the same region as the O—H band; however, further reaction causes no further changes in XPS and ATR IR spectra, suggesting that the

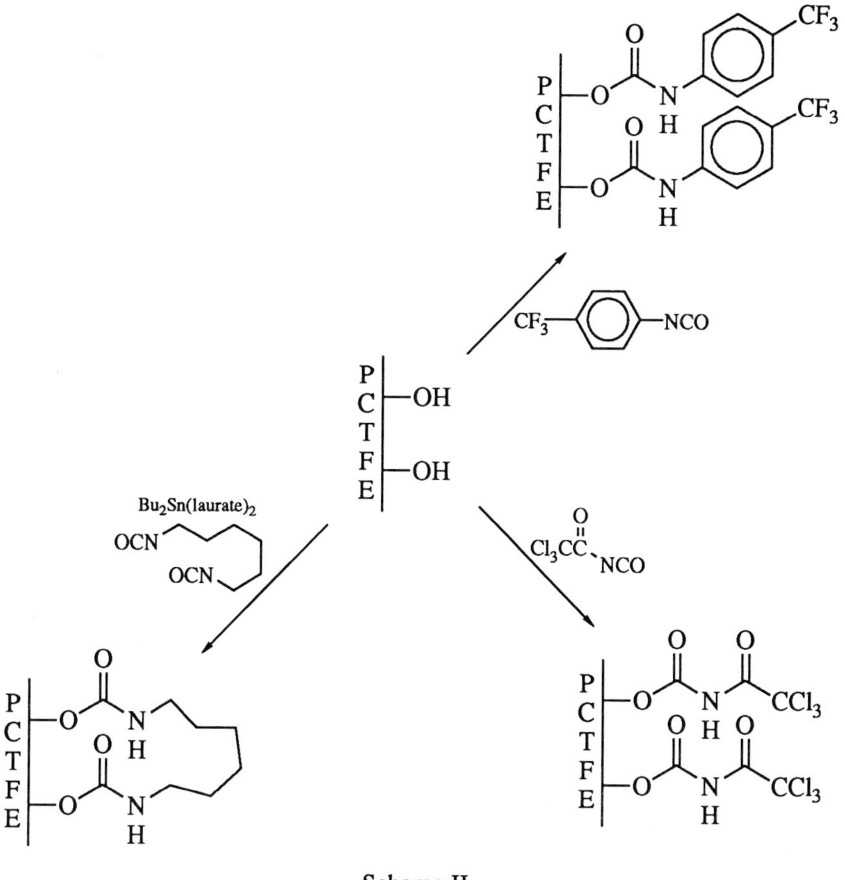

Scheme II

Polymer Surfaces and Interfaces II

Table 2. Water contact angle data for PCTFE–(CH$_2$)$_3$OH derivatives

Sample	Contact angles (deg)	
	θ_A	θ_R
PCTFE–(CH)$_3$OC(O)CH$_3$	75	35
PCTFE–(CH)$_3$OC(O)CCl$_3$	88	43
PCTFE–[(CH)$_3$OC(O)CH$_2$CH$_2$]$_2$	87	67
PCTFE–(CH)$_3$OC(O)CC$_6$F$_5$	94	62
PCTFE–(CH$_2$)$_3$OC(O)NHC$_6$H$_4$-p-CF$_3$	83	50
PCTFE–(CH$_2$)$_3$OC(O)NHC(O)CCl$_3$	77	48
PCTFE–[OC(O)NHCH$_2$CH$_2$CH$_2$]$_2$	73	38

reactions are complete. Notable absorbance bands are the two carbonyl bands (1744 cm^{-1} free, 1719 cm^{-1} hydrogen-bonded) and two aromatic C=C stretching bands (1618, 1607 cm^{-1}) in PCTFE–(CH$_2$)$_3$OC(O)NHC$_6$H$_4$-p-CF$_3$, the two free carbonyl bands (1806, 1746 cm^{-1}) and the two hydrogen-bonded carbonyl bands (1796, 1730 cm^{-1}) in PCTFE–(CH$_2$)$_3$OC(O)NHC(O)CCl$_3$ and

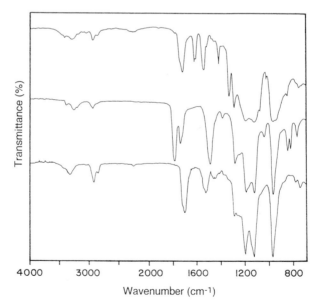

Figure 12. ATR IT spectra of (from top to bottom): PCTFE–(CH$_2$)$_3$OC(O)NHC$_6$H$_4$-p-CF$_3$, PCTFE–(CH$_2$)$_3$OC(O)NHC(O)CCl$_3$ and PCTFE–[(CH$_2$)$_3$OC(O)NHCH$_2$CH$_2$CH$_2$]$_2$. (Reprinted with permission from Lea and McCarthy, *Macromolecules*, **21**, 2318. Copyright (1988) American Chemical Society)

a weak $N{=}C{=}O$ stretching band (2272 cm^{-1}) (the weak intensity of the $N{=}C{=}O$ band indicates that half isocyanate/half urethane is present, but is a small contributor) and the two strong carbonyl bands ($1723, 1705 \text{ cm}^{-1}$) in PCTFE–[OC(O)NHCH$_2CH_2CH_2$]$_2$. The C 1s regions of these spectra along with fitted curves and their assignments are shown in Figure 13. The area data agrees well with the proposed structures in each case.

Urethanation chemistry was used for the synthesis of a polymer film surface which would form a strong adhesive bond to glass.[10] The reaction of PCTFE–(CH$_2$)$_3$OH with 3-isocyanatopropyltriethyoxysilane in the presence of dibutyltin dilaurate (a urethanation catalyst) in THF at room temperature for 24 hours yields PCTFE–OC(O)NH(CH$_2$)$_3$Si(OEt)$_3$:

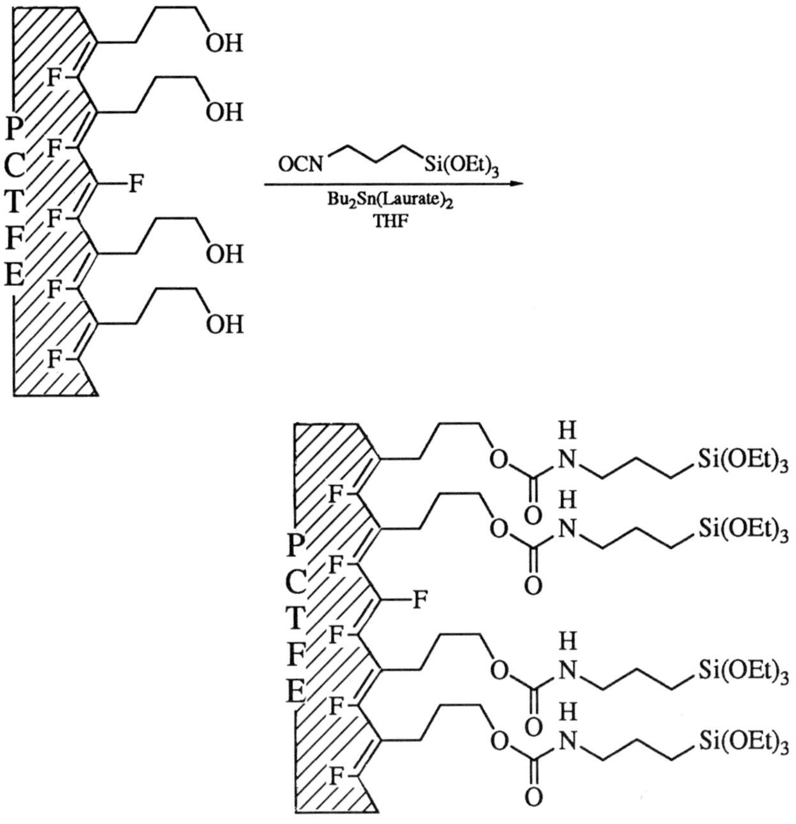

XPS and ATR IR indicate that the urethane is the sole product of this reaction. This modified film adheres tenaciously to glass when bonding is carried out at 80°C. The film cannot be removed from the glass without tearing it, indicating cohesive failure in the polymer film.

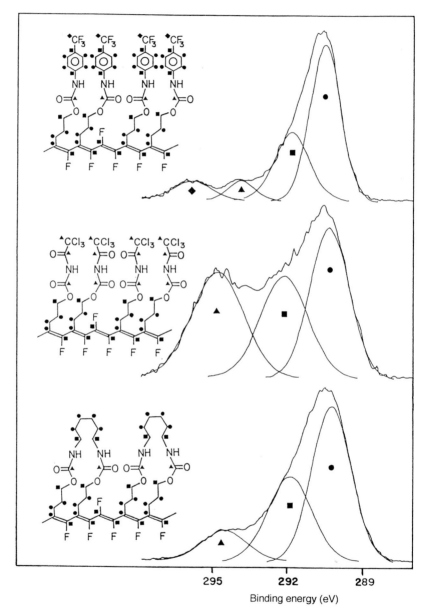

Figure 13. Carbon 1s regions of XPS spectra of (from top to bottom): PCTFE–(CH$_2$)$_3$ OC(O)NHC$_6$H$_4$-*p*-CF$_3$, PCTFE–(CH$_2$)$_3$OC(O)NHC(O)CCl$_3$ and PCTFE–[(CH$_2$)$_3$-OC(O)NHCH$_2$CH$_2$CH$_2$]$_2$. (Reprinted with permission from Lea and McCarthy, *Macromolecules*, **21**, 2318. Copyright (1988) American Chemical Society)

3 'SHORTCUTS' TO CARBOXYLIC ACID FUNCTIONALIZED SURFACES

Oxidant-sensitive polymers can be conveniently surface-functionalized with carboxylic acids by reaction with strong oxidizing agents. Chains are cleaved, carbon dioxide is produced and chain ends are left functionalized with carboxylic acids (the highest oxidation state of attached carbon). The fluoropolymers PVF_2, FEP and PCTFE are inert to strong oxidants and thus cannot be functionalized by oxidation. We have developed[11] two-step modification procedures to introduce submonolayer levels of carboxylic acid groups to these polymers using the following strategy:

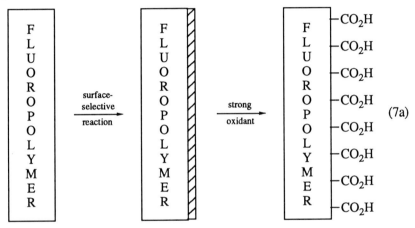

(7a)

Surface-selective reactions are carried out to convert a thin layer of fluoropolymer (several tens of angstroms) to an oxidant-sensitive product. The modified product is then completely removed by oxidation; the fluoropolymers are inert to these oxidation conditions. At sites where virgin polymer and the modified layer were covalently attached (partially modified chains), carboxylic acids are produced as chain termini (the following equation describes this for dehydrofluorinated PVF_2):

$$\text{\textasciitilde CH}_2\text{CF}_2\text{CH}_2\text{CF}_2 \quad \longrightarrow \quad \text{\textasciitilde CH}_2\text{CF}_2\text{CH}_2\text{CF}_2\text{-CO}_2\text{H} \quad (7b)$$

This system is unique because the oxidation stops when virgin polymer is reached and the topography of the surface is controlled and can be controlled by the sharpness/diffuseness of the modified polymer–virgin polymer interface. We studied butyllithium—modified PCTFE (equation 1), sodium naphthalide—reduced FEP[12] and dehydrofluorinated PVF_2. We review here the PVF_2 modification.

PVF_2 was dehydrofluorinated[2] with aqueous sodium hydroxide and tetrabutylammonium bromide at 40°C. The thin modified layer (PVF_2–CH=CF—) was removed oxidatively using potassium chlorate/sulfuric acid to yield PVF_2–CO_2H. Water contact angle data for these surfaces are given in Table 3. ATR IR and SEM of film samples showed no changes upon dehydrofluorination or oxidation. The C 1s regions of the XPS spectra and the UV–vis spectra of PVF_2, PVF_2–CH=CF— and PVF_2–CO_2H are displayed in Figure 14. Elimination renders a broad absorbance (conjugated —CH=CF—) and a decrease in the intensity of the CF_2 photoelectron line. Upon oxidation, the UV-vis and C 1s spectra of PVF_2–CO_2H become indistinguishable from those of PVF_2. The only difference in the XPS spectra of PVF_2 and PVF_2–CO_2H is the presence of an oxygen 1s line at 536 eV. XPS atomic composition indicates an approximate structure for the outer 10 Å of $(CH_2CF_2)_{13}CO_2H$, or one carboxylic acid per thirteen repeat units of PVF_2. That the oxygen introduced is indeed present as carboxylic acids was demonstrated by labelling with thallium ethoxide, reduction to the alcohol with borane and measuring the pH dependence of the water contact angles. Figure 15 shows the pH-dependence of the advancing and receding contact angles for PVF_2, PVF_2–CO_2H and PVF_2–CH_2OH (the borane-reduced product). Only PVF_2–CO_2H shows pH-dependent advancing and receding contact angles with $\theta_A/\theta_R = 77°/39°$ at low pH and $\theta_A/\theta_R = 68°/25°$ at high pH. This indicates that the surface becomes more hydrophilic when PVF_2–CO_2H is titrated to PVF_2–CO_2^- and that the surface carboxylic acids have pK_a values ranging from ~5 to ~10. Similar behavior is observed for oxidized polyethylene (PE–CO_2H).[13]

Table 3. Water contact angle data for PVF_2, PEEK and FEP derivatives

Sample	Contact angles (deg)	
	θ_A	θ_R
PVF_2	104	77
PVF_2–CO_2H	88	43
PEEK	85	55
PEEK ~ C=N—NHϕ(NO_2)_2	78	47
PEEK ~ C=CH_2	78	38
PEEK ~ CBrCH_21Br	87	56
PEEK ~ CHOH	69	23
PEEK ~ CHCl	79	47
FEP	115	100
FEP–PLL	80	16

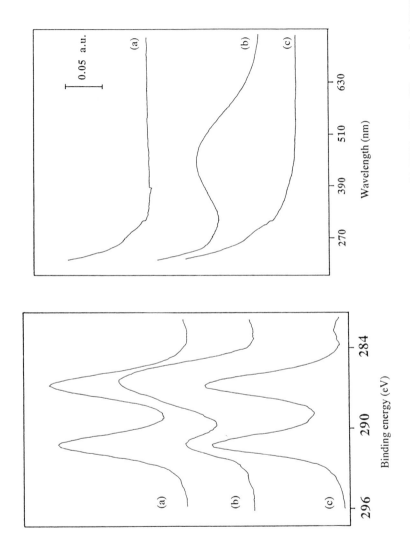

Figure 14. XPS (C 1s region, 15° takeoff angle) and UV–vis spectra of (a) PVF_2, (b) PVF_2–CH=CF— and (c) PVF_2–CO_2H. (Reprinted with permission from Shoichet and McCarthy, *Macromolecules*, **24**, 982. Copyright (1991) American Chemical Society)

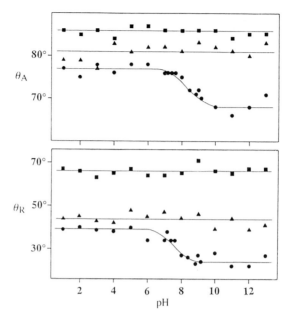

Figure 15. Dependence of θ_A and θ_R on pH (buffered aqueous solutions): ■ = PVF$_2$, ▲ = PVF$_2$–CH$_2$OH, ● = PVF$_2$–CO$_2$H. (Reprinted with permission from Shoichet and McCarthy, *Macromolecules*, **24**, 982. Copyright (1991) American Chemical Society)

4 POLY(ETHER ETHER KETONE) (PEEK)

PEEK is an engineering thermoplastic with excellent mechanical properties and a variety of materials uses. Due to its crystallinity, it is resistant to solvents and chemicals. As a chemically resistant film substrate it is particularly attractive as it contains a potentially reactive functional group, the diaryl ketone. We have carried out a number of ketone-selective reactions[14] on this film sample; several are described here.

PEEK

PEEK reacts with 2,4-dinitrophenyl hydrazine in acidic THF at 45°C:

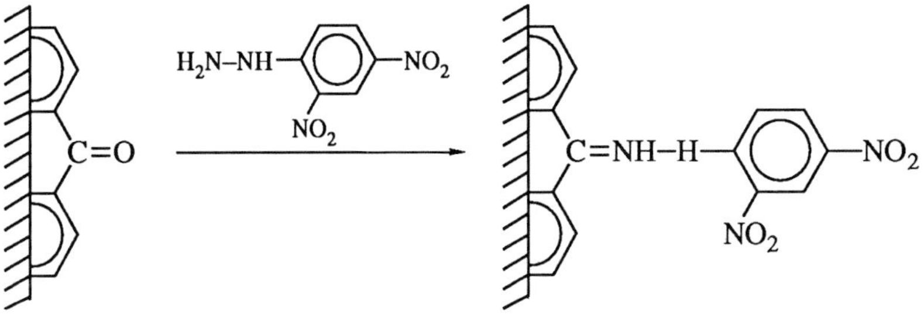

(8)

The resulting film product, PEEK $\sim$ C=N—NHϕ(NO$_2$)$_2$, exhibits nitrogen 1s peaks in the XPS spectrum at 406 eV (NO$_2$) and 400 eV (C=N—NH—). The repeat unit of the expected product predicts an XPS composition of C$_{25}$O$_6$N$_4$; C$_{25}$O$_{6.4}$N$_{2.1}$ is observed indicating a $\sim$50% yield of hydrazone. The diarylketone was converted to an olefin using Wittig conditions (reaction with methylenetriphenylphosphorane):

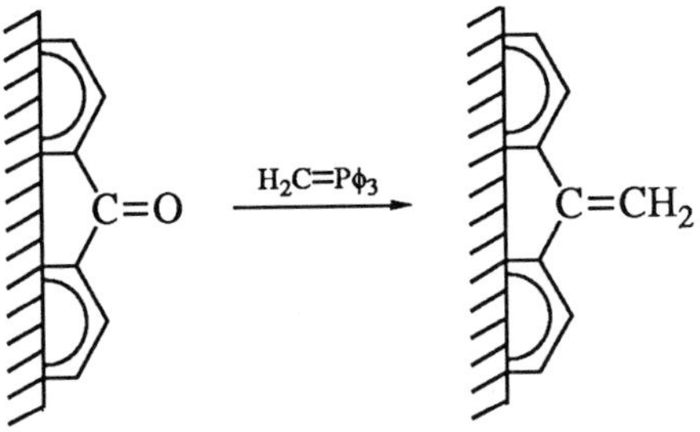

(9)

The observed XPS composition is C$_{20}$O$_{2.5}$, compared with the expected surface composition for a quantitative yield of C$_{20}$O$_2$, indicating a $\sim$50% yield. Reaction of PEEK $\sim$ C=CH$_2$ with bromine to form PEEK $\sim$ CBrCH$_2$Br confirms the presence of unsaturation. The bromination occurs in good yield (85% based on the 50% yield olefination), giving a surface with a structure of C$_{39}$O$_{5.0}$Br$_{1.7}$. PEEK does not react with bromine under the same conditions. Diarylketones at the PEEK surface are reduced to secondary alcohols (PEEK $\sim$ CHOH) using sodium bis(2-methoxyethoxy)-aluminum hydride;

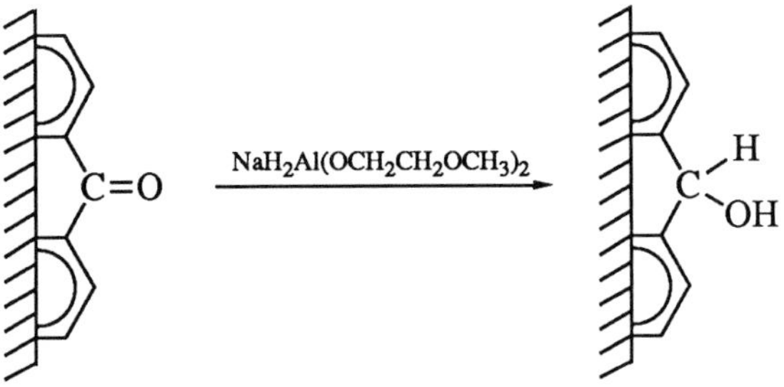

(10)

Analysis of the O 1s spectrum indicates a 90% conversion of the outer 10 Å and a 75% conversion of the outer 40 Å. These yields are calculated using the O 1s spectra. The O 1s region of PEEK exhibits two partially resolved lines, a low binding energy peak due to the ketone and a high binding energy peak due to the ether, in the expected 1:2 intensity ratio. Upon reduction this ratio increases. PEEK ~ CHOH is converted to the chloride using thionyl chloride:

$$(11)$$

XPS indicates that the reaction proceeds in ~75% yield. Water contact angle data for PEEK derivatives are given in Table 3.

5　ADSORPTION OF POLY(L-LYSINE) TO FEP

An alternative approach to introducing functional groups to chemically resistant polymers is by adsorption of functionalized polymers to the surface. We have been studying[15] the adsorption of poly(L-lysine) (PLL) from buffered aqueous solutions and alcohol/water mixtures to FEP surfaces. The driving force is the decrease in interfacial free energy between FEP and the solution. We describe here just one example. When FEP film is placed in a pH 11 solution of PLL ($M_w = 400$ K, 0.1 g/L) at 25°C, PLL adsorbs to the water–FEP interface irreversibly. It cannot be removed from the FEP surface by washing with pH 11 buffer or water. Water contact angle data for this experiment are included in Table 3. Figure 16 shows XPS spectra of virgin FEP and this FEP–PLL sample. The FEP–PLL spectra show, in addition to the FEP F 1s and high binding energy C 1s photoelectron lines, O 1s (536 eV), N 1s (403 eV) and low binding energy C 1s peaks, indicating the presence of poly(L-lysine). The reactivity of the ε-NH$_2$ groups in FEP–PLL was determined by reaction with 3,5-dinitrobenzoylchloride:

$$(12)$$

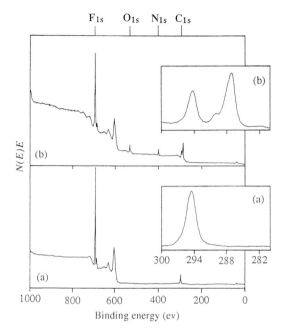

Figure 16. XPS survey and C 1s region spectra of (a) FEP and (b) FEP–PLL. (Reprinted with permission from Shoichet and McCarthy, *Macromolecules*, **24**, 1441. Copyright (1991) American Chemical Society)

XPS analysis of the N 1s region showed a $-NO_2$ (412 eV): $-NHC(O)-$ (405 eV) peak ratio of (0.67), indicating a $\sim 67\%$ reaction yield.

REFERENCES

1. Lee, K.-W., and McCarthy, T. J., *Macromolecules*, **21**, 309 (1988).
2. Dias, A. J., and McCarthy, T. J., *Macromolecules*, **17**, 2529 (1984).
3. Rasmussen, J. R., Stedronsky, E. R., and Whitesides, G. M., *J. Am. Chem. Soc.*, **99**, 4736 (1977).
4. Rasmussen, J. R., Bergbreiter, D. E., and Whitesides, G. M., *J. Am. Chem. Soc.*, **99**, 4746 (1977).
5. Cross, E. M., and McCarthy, T. J., *Macromolecules*, **23**, 3916 (1990).
6. Dias, A. J., and McCarthy, T. J., *Macromolecules*, **18**, 1826 (1985).
7. Bee, T. G., and McCarthy, T. J., submitted for publication.
8. Kolb, B. U., Patton, P. A., and McCarthy, T. J., *Macromolecules*, **23**, 370 (1990).
9. Lee, K.-W., and McCarthy, T. J., *Macromolecules*, **21**, 2318 (1988).
10. Lee, K.-W., and McCarthy, T. J., *Macromolecules*, **21**, 3353 (1988).
11. Shoichet, M. S., and McCarthy, T. J., *Macromolecules*, **24**, 982 (1991).
12. Bening, R. C., and McCarthy, T. J., *Macromolecules*, **23**, 2648 (1990).
13. Holmes-Farley, S. R., Reamey, R. H., McCarthy, T. J., Deutch, J., and Whitesides, G. M., *Langmuir*, **1**, 725 (1985).
14. Franchina, N. L., and McCarthy, T. J., *Macromolecules*, **24**, 3045 (1991).
15. Shoichet, M. S., and McCarthy, T. J., *Macromolecules*, **24**, 1441 (1991).

2

Self-assembled Molecular Films as Polymer Surface Models

D. L. Allara, S. V. Atre and A. N. Parikh

*Department of Materials Science
and Department of Chemistry
Pennsylvania State University*

1 INTRODUCTION

1.1 The Goal: Structure–Property Relationships for Bulk Polymer Surfaces

In many cases the performance of a polymer item is dominated by the surface properties of the polymer. Examples of such surface properties are: wettability, friction, lubricity, wearability, chemical reactivity, biological compatibility and activity, permeability, charge storage capacity and electrical response. These, and other, surface properties depend critically upon the chemical and physical details of molecular structure at the surface of the polymer.[1] The important chemical characteristics are derived from the type of exposed chemical functionality and their associated features: lateral distribution including surface density and translational ordering; bond orientations with respect to the surface plane; composition and surface distribution for multiple surface functionality; and the depth distribution of functionality and related features, viz. below the surface and into the polymer interior. The important physical characteristics include roughness, void content, mass density and defects. In order to obtain the capability to control surface properties by manipulating surface structure it is necessary to have an extensive database of detailed correlations between properties and structure for the polymer surfaces of interest. However, other than generalizations about simple behaviors, e.g. wetting and chemical reactivity, very little definitive work has been reported on such structure–property correlations.

Polymer Surfaces and Interfaces II
Edited by W. J. Feast, H. S. Munro and R. W. Richards
© 1993 John Wiley & Sons Ltd

1.2   The Problems: Characterization and Control of Polymer Surface Structures

There are two major reasons for the above lack of definitive surface–property correlations. First, polymer surface structures are very difficult to control and modify in a systematic way in order to emphasize selected features of interest. Second, the subtle but important features of interest are extremely difficult to characterize at the level of detail necessary to correlate accurately with surface properties, particularly complex ones such as biocompatability.

The first reason above stems from the fact that the processing of a polymer to provide a workable surface, e.g. hot pressing, usually produces just one kind of surface structure with little controllable variability possible. In addition, the processing often introduces unwanted and uncontrolled impurities or damage, e.g. mold release agents, surface oxidation or roughness in the case of hot pressing. Although in principle these may be the 'real world' surfaces, in practice it is difficult to reproduce such surfaces to make sense out of a body of surface property measurements.

The second reason, difficulty with characterization, stems from the fact that in spite of the level of sophistication of many surface science tools, only a few are well suited for characterizing the necessary detail of features in polymer surfaces. The requirements are rather severe. First, the characterization probe must be nondestructive. This is a minimal requirement for stable inorganic materials, but polymers, as organic materials, are often unstable with respect to electron, ion and intense laser beams as well as heat. The latter is particularly relevant since polymer surfaces often have very dynamic structures which are temperature and solvent dependent. Unlike inorganic materials, polymers rarely form good single crystals and cannot be sputtered and annealed to produce initially clean, pure surfaces even vaguely representative of intrinsic material. In addition, for polymers the probe technique must be sensitive to molecular features such as conformational sequences and hydrogen bonding. Since changes in these features are defined by subtle differences in energy levels of the structures, highly sensitive valence band or vibrational spectra are needed for characterization. Finally, polymers may need to be characterized directly and in real time in the presence of the liquid environment to capture features such as solvent and time-dependent confirmations at the liquid–solid interface. This is particularly true for biological interfaces, which need to be characterized for biomedical materials applications. Unfortunately, most typical surface science techniques are operable only with the sample under vacuum.

In spite of the above requirements there is a growing number of techniques that have been developed to aid efforts in polymer surface characterization. Table 1 gives a selection of those techniques which are sensitive to molecular features (and thus nondestructive) and also lists attributes of the techniques with respect to important characterization needs. An important point to note is that several techniques that are *not* surface selective, infrared spectroscopy

(IRS), spectroscopic ellipsometry (SE) and Raman, are very powerful structural tools and possess the capability to directly probe the liquid-solid interface. However, these techniques are only infrequently used for polymer surface analysis because of the lack of surface selectivity. Further, IRS and SE can yield very quantitative information on a millisecond to second time scale, a scale very useful for following many surface processes directly, particularly chemical changes. One should note that these techniques all use photons as probes.

The character of the surface selectivity of a specific analysis technique is crucial in determining the optimum use of the technique. It is reasonable to assume that almost all the important surface information is contained in the first 5 nm of a polymer surface. For example, wettability and biocompatability are generally considered to be controlled by the top 5–10 Å of the surface. Table 2 lists a variety of molecular structure-sensitive techniques and characterizes them in terms of their abilities to analyze the top 5 nm or less of a polymer film. Table 2 also considers two important sample characteristics, surface roughness and inherent bulk–surface contrast, which exert a profound effect on the tractability of surface analysis. The presence of surface roughness adds complexity to the analysis of a measurement and in most cases tends to lower the information content. The case of low bulk–surface contrast is a common one in which the molecular nature of the surface differs only slightly from that of the bulk. In this case, nonsurface selective measurements yield information about the surface region which is obscured by a large bulk signal nearly identical to that from the surface.

One fact is very obvious. As the film thickness decreases to form a smooth, planar monolayer, the number of applicable techniques for surface analysis increases significantly. This point is illustrated qualitatively in Figure 1 where the authors' rough estimate of the useful surface information content of a measurement is plotted against the sample thickness of a film on a smooth, planar substrate. The three techniques shown, IRS, SIMS (secondary ion mass spectrometry) and XPS (X-ray photoelectron spectroscopy), are judged by these authors to be the most all-round, useful, cost effective techniques for determining structure in polymer films.

1.3   One Solution: Controlled Structure Self-assembled Monolayers as Model Polymer Surfaces

The previous section establishes the case that monolayer thickness films on smooth, planar substrates are amenable to the broadest range of useful analysis techniques for probing the sample surface. A further point is that thin films can often be prepared with extremely smooth surfaces, ideal for quantitative surface analysis. These points are summarized in Table 3 in which the picture at the top illustrates the basis for the concept of using thin films and monolayers as polymer surface models. The use of a molecular monolayer in this regard seems

Table 1. Attributes of molecular sensitive[a] techniques applicable to the characterization of polymer surfaces

Analysis technique	Quantifiable	Lateral imaging	Depth profiling (nondestructive)	Surface or interface selection	In situ capability	Measurement time scale (for monolayer)	Reference
X-ray photoelectron spectroscopy (XPS)	+	≥10 μm	≥5 Å	+	−	≥1 s	2
Secondary ion mass spectrometry (SIMS)	−	≥0.1 μm	−	+	−		3,4
Electron energy loss spectroscopy (EELS)	−	−	−	+	−		5
Atomic force microscopy (AFM)	+	Atomic	−	+	+	≥1 s	6

Technique							
Sum frequency generation (SFG)	+	–	–	+	+	≥10 ps	7, 8
Near edge X-ray absorption fine structure (NEXAFS)	–	–	–	+	–		9
Infrared spectroscopy (IRS)	+	≥10 μm	≥20 nm[b]	–	+	≥100 ms	10, 11
Spectroscopic ellipsometry (SE)	+	–	≥0.1 nm	–	+	≥50 ms	12
Raman	–	≥1 μm	–	–	+	Slow	13–15

[a] Molecular sensitive implies responsive to one or more molecular features including chemical bonding, valence states, coordination symmetry, bond and group orientation, bond conformations and intra- and intermolecular interactions (e.g. hydrogen bonding).
[b] Depth profiling is appropriate only for internal reflection spectroscopy in which the internal reflection element is in intimate contact with the polymer surface.

Table 2. Applicability[a] of technique to analyze a ≤5 nm surface region for different sample types

Technique[b]	Bulk polymer			Smooth, thin (<50 nm) polymer film on a planar substrate	Molecular monolayer surface on a planar substrate
	Rough, non-planar surface; bulk and surface very similar	Highly smooth, planar surface; bulk and surface very similar	Highly smooth, planar surface; bulk and surface very different		
XPS (core)	+	+	+	+	+
XPS/UPS (VB)	+	+	+	+	+
SIMS	+	+	+	+	+
EELS	+	+	+	+	+
NEXAFS	+	+	+	+	+
AFM	+	+	+	+	+
ISS		+	+	+	+
RBS/FBS			+	+	+
IRS			+	+	+
SE				+	+
Raman				+	+
NR			+	+	+
XRR				+	+
GIXRD				+	+
He scattering					+
LEED					+
STM					+

[a] Qualitative assessment of general applicability. In some cases extraordinary circumstances may allow an analysis where no + is given.
[b] Definition of technique acronyms:
Spectroscopies: XPS, X-ray photoelectron; UPS, ultraviolet photoelectrons; SIMS, secondary ion mass; EELS, electron energy loss; NEXAFS, near edge X-ray adsorption fine structure; ISS, ion scattering; RBS/FRS, Rutherford back scattering and forward recoil scattering; IRS, infrared; SE, ellipsometry.
AFM, atomic force microscopy; NR, neutron reflectometry; XRR, X-ray reflectometry; GIXRD, grazing incidence X-ray diffraction; LEED, low-energy electron diffractions; STM scanning tunneling microscopy.

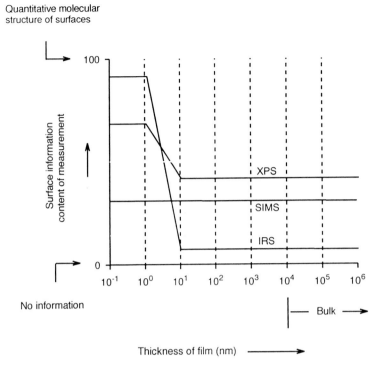

Figure 1. A schematic representation of the approximate relationship between the level of surface information obtained from a given analysis as a function of the film thickness

extremely attractive. Recently there have been great advances in the development of self-assembled monolayer structures from different types of molecules adsorbed at several types of substrates.[16] In some cases films have been produced with ordered arrays of surface groups. A strategy of how to go about synthesizing an appropriate polymer surface model from a monolayer structure is indicated in Figure 2. Here the concept of a peg-board model of the polymer surface is introduced. The latter, in general, will consist of folds and chain ends, exiting and reentering a crystalline or amorphous matrix. If the outer surface were hypothetically sliced off one would see an array of the cross sections of the sliced chains as a lattice whose patterns would need to be copied as the pinning or attachment lattice for a self-assembled monolayer. Although a complicated polymer surface pattern could never be replicated precisely, the general character could be copied. For example, a non-uniform lattice could be approximated as random. In cases of regular lattice structure, close approximation seems possible, e.g. a regular hexagonal lattice with a given surface density of points. It is clear that the success of this peg-board strategy depends

Table 3. Selected characteristics of preparation and analysis of bulk polymer surfaces, thin polymer films and molecular monolayers

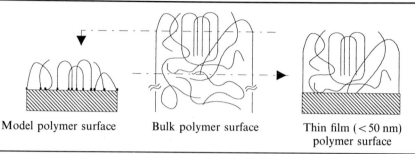

Model polymer surface	Bulk polymer surface	Thin film (<50 nm) polymer surface
1. Surface layer can be quantitatively characterized: (a) Nano-scale thickness (b) Smooth, planar (can be atomically smooth) substrate	1. Surface difficult to characterize: (a) Bulk present (surface sensitivity necessary)	1. Surface layer can be qualitatively characterized, occasionally quantitatively: (a) Bulk minimized (b) Topography smooth
2. Surface layer can be prepared in precise manner: (a) Full range of surface chemistry knowledge and techniques available (b) Effort must be applied to capture real polymer surface features	2. Surface difficult to prepare in controlled manner: (a) Range of polymer variations and processing very limited	2. Surface still difficult to prepare in a controlled manner: (a) Range of polymer variations and processing very limited (easily cast films required)

upon the experimental capability to assemble flexible chain molecules into selected patterns of chain ends and folds at surfaces. Further, the molecular assembly must be sufficiently thick that the surface plane will be $\gtrsim 1$ nm away from the substrate surface plane, a distance needed to guarantee that the general sample surface properties, such as wetting, are unaffected by the substrate character. In view of the rapid advances that have been made recently in the area of self-assembled monolayers it appears that the peg-board modeling strategy is a workable one within the limitations imposed by the range of self-assembled monolayer structures accessible by fortuitous combinations of surface chemistry and intrinsic substrate surface lattice arrangements. In the next section we describe recent examples from our own work in which polymer surface models have been synthesized using the self-assembled techniques.

Bulk polymer surface

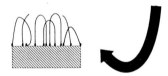

Peg-board model: polymer surface

Self-assembled monolayer

Figure 2. A conceptual strategy for building model polymer surfaces

2 STRATEGIES FOR USING SELF-ASSEMBLED MONOLAYER FILMS AS POLYMER SURFACE MODELS

2.1 Requirements for Polymer Surface Models

In order to be the most useful as polymer surface models, self-assembled monolayer structures should meet the following ideal requirements: (a) highly smooth and planar supporting substrates, (b) complete coverage of the substrate surface by the film such that substrate effects are not present in the monolayer surface property measurements, e.g. wetting, (c) wide variability of selection of the outer (ambient) surface functionality in terms of chemical types, including mixed compositions, (d) flexibility in control of surface density, lateral distribution patterns and vertical placement ('pockets' and 'hillocks', similar to biological receptor sites) and (e) capability to control the surface population of chain folds and chain ends. Fulfilment of the above requirements will allow maximum control of the surface structure and the maximum degree of characterization, e.g. by use of the tools listed in Table 2. Of course, at the present state of the art, such ideal control is not possible, but many requirements, in fact, can be reasonably met.

2.2   The Different Types of Self-assembled Monolayers

Monolayers can be formed by a variety of methods including vacuum evapora-tion and Langmuir–Blodgett deposition. The self-assembly process refers spe-cifically to a monolayer that is spontaneously formed during the contact of a fluid phase of molecules with a surface.[16] The fluid phase consists of a gas, a solution or a pure liquid. The molecules can consist of simple ones as well as polymers, but must contain at least one functional group that is capable of spontaneous chemisorption at specific sites on the substrate surface. Because of this specific molecular group–surface site reaction, additional layers cannot form except by fortuitous molecule–molecule interactions or chemical reactions. The substrates can consist of a wide range of solid state materials, usually inorganic, with reactive surface sites arranged in both random distributions, e.g. amorphous oxides, and highly ordered distributions, e.g. single-crystal semiconductors. It is rather obvious that the most ordered self-assembled monolayer structures should be formed from appropriately sized and shaped molecules attached firmly at regularly spaced lattice points on a uniform single-crystal surface. These types of ordered monolayers are readily formed by the chemisorption of n-alkane thiol molecules on single-crystal gold surfaces. For example, on a (111) surface the thiol groups are pinned on a $\sqrt{3} \times \sqrt{3}$ R30° lattice with a 4.9 Å spacing and the rod-like chains are tilted at an angle of 26–28° from the surface normal.[17] Less obvious is the fact that ordered films can be formed when the molecule–molecule packing forces drive the organiza-tion while the molecule–substrate interactions are compliant to this packing arrangement. Examples of such ordered films are: (a) hydrolyzed octadecyl trichlorosilane on Au(111),[18] where the crosslinked planar siloxy network holds the assembly to the gold by purely uniform, nonlocalized dispersion forces and (b) n-alkanoic acids chemisorbed on the amorphous, poorly defined oxide 'monolayers' of evaporated silver as ionic carboxylate species, which apparently are free to arrange themselves according to the lowest chain-packing energy of an orthorhombic type of unit subcell for the alkyl chains.[19]

The chemical nature of the substrate–functional group attachment interaction is very specific and quite important in designing strategies to self-assemble multifunctional molecules to give specific surface structures and properties. If additional substrate-reactive functionality is present in an adsorbate molecule and the molecular structure is sufficiently compliant, e.g. a flexible chain, then the molecule can form folded or bent structures. Such would be the case especially for polymeric molecules with side chain functionality. To ensure, on the other hand, that additional functional groups are not chemisorbed, then they must be selected such that they do not possess inherent chemical reactivity with the substrate or that they are sterically blocked from reacting. A typical strategy is to select groups with intrinsically different hard–soft donor acceptor properties from those of the adsorbate. Such is the strategy used to produce the very popular terminally functionalized monolayers of $X(CH_2)_nS$—moieties on Au(111).[20–23] The polarizable thiolate group binds to the gold by strong soft–soft donor acceptor interactions which are overwhelmingly selective in adsorption competition with a variety of harder, less polarizable groups such as OH, CO_2H, NH_2, CO_2CH_3, etc. This strategy gives very uniform highly organized surface arrays of common functional groups. These types of surfaces are extremely useful for constructing idealized polymer surface models.

3 RECENT RESULTS IN THE SYNTHESIS, CHARACTERIZATION AND SURFACE PROPERTIES OF SELF-ASSEMBLED MONOLAYERS

In this section we discuss recent results, primarily from our laboratory, in the preparation of specific types of surface structures via molecular assembly. The surface structures studied fall into three general classes: (a) derivatized macro-molecular monolayer films, (b) short flexible chains attached to the surface at one chain end and (c) short flexible chains attached to the surface at both chain ends. The primary surface property of interest will be liquid wetting. In all cases, because of both the monolayer-sized thickness of the films and the highly smooth, planar substrates, quantitative analysis by a variety of techniques has been possible. In particular, IRS, XPS and SE have been extremely useful.

3.1 Derivatized Polyacrylic Acid Monolayers on Oxidized Aluminum

Polyacrylic acid (PAA) readily adsorbs from ethanol solution on to the intrinsically basic surface of the hydrated native oxide of aluminum to form a monolayer-sized thickness film (~ 8–10 Å) which appears to uniformly cover the surface in the form of a quasi two-dimensional layer of strings randomly pinned on the surface.[24] Analysis of the IRS spectrum reveals that both undissociated CO_2H groups and $CO_2{}^-$ groups are present and that their ratio is approximately 1:2. Based on the extensive knowledge of the behavior of CO_2H groups at Al_2O_3 surfaces[25,26] it is a straightforward conclusion that the polymer is bound to the surface by the $CO_2{}^-$ groups. The remaining $\sim 1/3$ of the original CO_2H groups remain protonated as 'free' acid. The accessible nature of the CO_2H groups can be easily demonstrated by derivatization with a number of reactions. In recent studies[27] we have been able to attach a variety of derivatives. Since the position of the CO_2H groups is random, the location of the derivative groups will accordingly be identically random. Figures 3 and

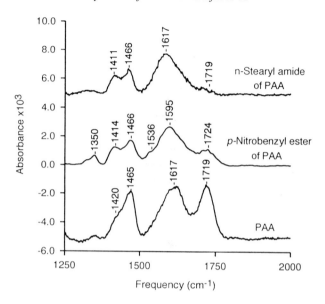

Figure 3. Low-frequency infrared spectra of functionalized and unfunctionalized PAA films on oxided aluminum substrates. The position of the baselines on the absorbance axis is arbitrary

4 show IRS spectra for PAA monolayers derivatized by the stearyl amide group, $C_{18}H_{37}NH\overset{O}{\overset{\|}{C}}$, and the *p*-nitrophenyl group, $O_2NC_6H_4O\overset{O}{\overset{\|}{C}}$. These spectra, when compared to the original PAA–Al_2O_3 spectrum, clearly show the presence of the derivative group. From the nearly complete loss of the CO_2H carbonyl stretching feature at ~ 1720 cm^{-1} one can calculate correspondingly high yields for the derivatization reactions. The XPS spectra are entirely consistent with the IRS evidence. Ellipsometry measurements show that the stearyl amide and *p*-nitrophenyl groups add ~ 5–10 Å to the thickness of the initial PAA films.

The wetting properties of the derivative films reflect the changes from the PAA film properties. The water contact angle (static pendant drop) variation as a function of pH is shown in Figure 5. Whereas the acidic PAA surface shows the expected dependence, the derivatized groups show pH independent behavior. These data show that a layer of thickness of ~ 5–10 Å is sufficient to completely dominate wetting behavior, a currently well-accepted fact. Applying Cassie's law[28] as a rough approximation to the case of the stearyl amide film, we calculate a coverage of ~ 30–40% of a complete monolayer of $C_{18}H_{37}$ chains packed at the pure liquid density. This agrees qualitatively with the ellipsometry-derived thickness as well as thicknesses estimated from IRS and XPS data. All the data on the stearyl amide monolayer, taken together, point to the fact

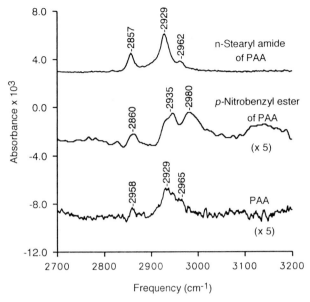

Figure 4. High-frequency infrared spectra of functionalized and unfunctionalized PAA films on oxided aluminum substrates. The position of the baselines on the absorbance axis is arbitrary. The two lower spectra have been scaled for convenience

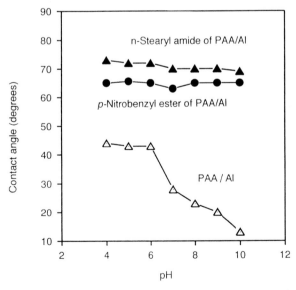

Figure 5. The dependence of water contact angles on drop pH for the surfaces of functionalized and unfunctionalized PAA films on oxided aluminum substrates

that a low-density, amorphous polyethylene-like surface has been synthesized. The advantage of this particular type of polymer surface model is that it is based on an actual polymer monolayer, robustly bound (it survives boiling water for at least one minute) to a smooth, planar surface. However, a disadvantage of this system is that the surface groups will always be randomly spaced and their surface density is fixed. The use of self-assembled monolayers of small molecules can lift this random 'pinning lattice' constraint, as will be shown in the subsequent sections.

3.2   Terminally Substituted n-Alkane Thiols on Au(111)

As mentioned in Section 3.2, adsorption of $X(CH_2)_nSH$ (or $[X(CH_2)_nS]_2$) on to Au(111) surfaces forms an ordered monolayer film, which for the proper choice of X leaves the exposed surface covered by a layer of X groups. Typical choices of X are OH, NH_2, CO_2H, CO_2CH_3, $CONH_2$ and $—OR$. Since the original discovery of this system,[20,21] an extensive amount of characterization of the wetting properties of both the pure X surfaces and mixed group surfaces has been reported.[22,23] The latter can be prepared by adsorption of mixtures of compounds with different terminal groups X and X' and different sizes of the chain lengths (values of *n*). These studies are fully documented in the literature so results will not be reviewed here. It is reasonable to assert that our quantitative understanding of the wetting properties of a variety of these groups and their combinations has been dramatically increased because of these model surface studies. Appropriate to the earlier discussion in this paper (Section 1) one of the major reasons for the utility of these surfaces has been the fact the structures can be characterized quantitatively[17,22] and thus quantitative structure–property correlations derived. An example of the detail of characterization possible is given for the monolayer made from $HS(CH_2)_{15}CO_2CH_3$. Ellipsometry shows this monolayer to be about 22 Å in thickness.[22] From recent infrared spectroscopy studies alkyl chains in such monolayers have been shown to consist of highly all-*trans* conformations with tilt angles about 26° from the surface normal.[17] Results from the earlier study indicate that the ester group appears to have the $C=O$ bond oriented somewhat close to parallel with the surface plane.[22] The XPS spectra are quantitatively consistent with the above structure.[22] Further, for the case of the $X=CH_3$ films, transmission electron diffraction shows the films to be crystalline[29] (see Section 2.2) and at low temperatures the helium scattering diffraction pattern shows the terminal methylene groups to be hexagonally ordered[30] while crystal field splitting of the CH_2 bending modes in the IRS spectra shows that there are two chains per unit subcell.[31] Also, recent X-ray in-plane diffraction data confirms the crystalline packing of the chains.[32]

From the above discussion it should be very clear that the two major advantages of the linear chain thiol monolayer on gold systems for structure–

wetting correlations are that a variety of functionalized surfaces can be prepared and, very importantly, that these surface structures can be quantitatively characterized by a number of surface science methods.

3.3 Folded, Interior-substituted, α,ω-Dialkanoic Acids on Silver

In spite of the striking variability of surfaces which can be prepared via the linear alkane thiol–gold route, one extremely important class of surfaces that is not accessible is that consisting of interior (as opposed to terminal) functionality such as $-O-$, $-CH_2-$, $-S-$, $-CONH-$, etc. This type of functionality is extremely important because most polymer surfaces consist primarily of chain folds and surface properties such as wetting, or biological activity cannot be properly understood or modeled unless these specific structures are considered.

One successful strategy for preparing such surfaces has been discovered recently in our laboratory. It appears that α,ω-dicarboxylic acids of the type $HO_2C-(CH_2)_m-G-(CH_2)_nCO_2H$ adsorb from solution on to oxided silver surfaces with both acid groups located at the silver surface, the alkyl chain consequently folded into a 'loop' and the outer (ambient) surface composed of the interior backbone groups of the chain folds. In the case of $n = m = 14$ and $G = -(CH_2)_2-$, the adsorption from tetrahydrofuran solution can lead to a ~ 20 Å thick crystalline monolayer structure of tight chain folds which exhibits surface wetting properties indistinguishable from a polyethylene surface.[33,34] In particular, this monolayer shows completely pH-independent contact angles for water and exhibits a critical surface tension of 32 dynes/cm, the same value as that for polyethylene.

Again, of particular importance is the fact that the structure of this surface can be very accurately characterized compared to what is possible for the surface of a bulk polyethylene sample. The IRS spectra of a film are shown in Figure 6. From these spectra one can derive the fact that the film is crystalline, contains tilted all-*trans* sections of alkyl chains and a conformationally disordered section of alkyl chain. For brevity, the details of the analyses are not given here but can be found elsewhere.[33,34] Substitution of the central $-(CH_2)_6-$ section of the chain by $-(CD_2)_6-$ results in the conclusion that the disordered segment of the loop chain is about 6 methylene units (an error of $\sim \pm 1$) long.

In recent experiments[35] the group G has been changed from $-(CH_2)_2-$ to

$$-O-, \quad -S-, \quad -\overset{\displaystyle O}{\underset{\displaystyle \|}{C}}- \quad \text{and} \quad -O-(CH_2)_2-O-.$$

All of these dicarboxylic acids form folded structures on silver with surface wetting properties derived from the presence of the group G. In addition, IRS spectra show that for $G = -O-$ and $-S-$, crystalline structures form. The ether folds are particularly important since these types of chain segments often appear at real polymer surfaces

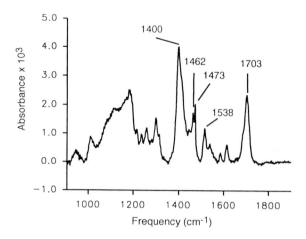

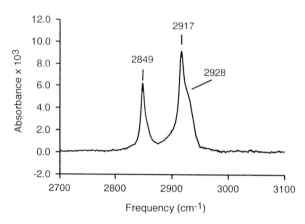

Figure 6. High- and low-frequency spectra of a monolayer of $HOOC(CH_2)_{30}COOH$ adsorbed on an oxided silver substrate

and dominate wetting properties. For example, for many polyurethane materials, the soft segments consist of ether groups derived from alkanediols and such segments inevitably migrate to the surface selectively compared to the polar urethane segments.[36] Further work in our laboratory is now also directed toward potentially biologically interesting surfaces such as those derived from polypeptide folds. Of course, a variety of combinations of other attachment molecules and substrates other than carboxylic acids on silver is possible. Alkane dithiols on gold is one system currently under investigation in our laboratory.[35]

3.4   An Alternative to RSH/Au: RSH/GaAs(100)

As mentioned previously (Section 3.2) self-assembled monolayers prepared from $X(CH_2)_nSH$ on Au(111) have become by far the most popular model systems for molecular surface properties. However, in spite of their enormous inherent chemical flexibility these films are limited in terms of the surface texture or corrugation. Although the surface can be textured vertically by preparing mixed composition films of different constituent chain lengths, the lateral spacing is fixed in the alkane thiol system to the $\sqrt{3} \times \sqrt{3}$ hexagonal lattice spacing as an intrinsic property of the Au(111) surface. In order to change the molecule–molecule spacing of the alkane chain adsorbates a different lattice pattern is needed. The square lattices of the Au(100) and Au(110) faces provide possible pinning lattices, but finding a range of other materials is of interest. On Ag(111) alkane thiols can be adsorbed on a hexagonal lattice with a spacing of ~ 4.4 Å,[17] very close to the closest packing chain–chain distance. However, because of the basic properties of native silver oxide, terminal functionality such as CO_2H tends to compete with the SH group for chemisorption and thus the preparative flexibility of the RSH–Ag system is comprised. Recently we have discovered that highly organized monolayers can be formed on GaAs(100) surfaces by simply heating the oxide-stripped substrate in concentrated solutions of $X(CH_2)_nSH$ compounds.[37] For the case of $X = CH_3$ and $n = 17$, IRS spectra, XPS spectra and ellipsometry all support a monolayer structure consisting of fully extended, all-*trans* chains, tilted at an angle of $\sim 57°$ from the surface normal and attached on a 5.6 Å square lattice as arsenic atoms, presumably via As—S bonds. Further details of the structure are not yet available, but the value of the chain tilt angle suggests that the chains are tilted along a direction close to the simple square cell diagonal, a distance of ~ 8 Å. From a simple analysis of molecular models it is clear that this distance requires a strong vertical modulation of the chain atoms along the surface with a considerable exposure of CH_2 groups, relative to the more nearly pure terminal group surfaces of the corresponding monolayers on Ag(111) and Au(111). A comparison of the liquid drop contact angle values for hexadecane (θ_{HD}) is consistent with this. For the CH_3 terminated surface prepared from $CH_3(CH_2)_{17}SH$, reported values of the contact angles[17,37] show θ_{HD} (Au(111)) $\cong \theta_{HD}$ (Ag(111)) = 48–50° and θ_{HD}(GaAs(100)) = 41°. For reference, θ_{HD} of a polyethylene surface (pure CH_2) = 0°. Although θ_{HD}(GaAs) is less than θ_{HD}(Au) by $\sim 8°$, the value might have been expected to be even lower, i.e. closer to θ_{HD} (pure CH_2), because of the large exposed CH_2 area in GaAs. These data are summarized in Table 4, which also shows schematic drawings of the chain structures illustrating the surface exposure of CH_2 in the highly tilted films. Whereas the quantitative details of these comparisons are not yet clear and require further experiments and analyses, it is clear that the ability to create a variety of model surface

Table 4. A comparison of wetting data from four different types of self-assembled monolayers of octadecyl chains exhibiting different chain tilt angles

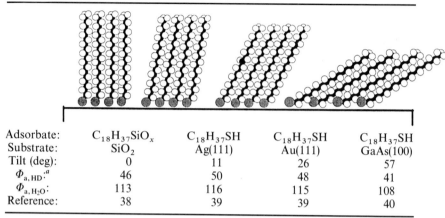

Adsorbate:	$C_{18}H_{37}SiO_x$	$C_{18}H_{37}SH$	$C_{18}H_{37}SH$	$C_{18}H_{37}SH$
Substrate:	SiO_2	Ag(111)	Au(111)	GaAs(100)
Tilt (deg):	0	11	26	57
$\Phi_{a,HD}$:[a]	46	50	48	41
Φ_{a,H_2O}:	113	116	115	108
Reference:	38	39	39	40

[a] Φ_a = advancing contact angle (degrees); HD = hexadecane.

structures opens up important new avenues to probe the urgent questions of the correlation between surface structure and wetting properties.

3.5   Disordered Surfaces

All of the above examples of surface structures have been drawn from systems of ordered self-assembly. In reality, the majority of polymer surfaces are disordered and models of these surfaces also need to be constructed and studied. Disordered surfaces are in some regards easier to make than ordered ones in that one can consider a poor preparation of an ordered surface, viz. a low coverage, or a large fraction of inert impurities, a disordered surface. However, in order to prepare a disordered surface in a rational way requires a specific strategy. A very straightforward strategy that we have recently used involves grafting alkyl chains on to a planar amorphous SiO_2 surface populated with ~ 5 silanol ($\equiv SiOH$) groups/nm² in a completely random distribution. Reaction of *n*-alkanols with the $\equiv SiOH$ groups at 140°C produces surface-bound silylalkyl ethers via dehydration.[41] The coverage of the alkyl chains grafted under the above conditions is ~ 3 chains/nm², as determined from ellipsometry, IRS and forward recoil scattering. The above value of the coverage is quite a reasonable one for a limit of steric crowding in which the chains can only half occupy that fraction of surface sites which are closer together than the effective chain hardwall diameter. This structure provides a model of a surface that is populated with a random distribution of alkyl chains and thus implies a molecular assembly with permanent translational disorder. However, there are two important features which could vary as a function of chain length: the extent of conformational disorder and the relative populations of chain folds and chain ends at the outer surface. These features are precisely the ones which can control polymer surface properties, and thus establishing model systems for their study is important. Their interplay with chain length and surface wetting can be seen in Figure 7 where the hexadecane contact angle is plotted against chain length. As the latter increases there is a corresponding increase in the oleophobicity of the surface. Since this series of films exhibits constant surface coverage with thicknesses sufficient to screen substrate effects on wetting (> 5–10 Å), the only changes possible to explain this behavior involve the variation of chain end population at the surface. Thus the wetting changes must arise from a rearrangement of the surface with increasing chain lengths to a structure with an increasing chain end contribution. This also implies a straightening of the chains to allow protrusion of the ends, a picture consistent with the IRS spectra which show decreasing conformational disorder with increasing chain length.[42]

Although the above system only allows study of $-CH_2-$ and $-CH_3$ groups it is obvious that more sophisticated film synthesis could be devised to include a variety of chain fold and chain end groups which then could be used to construct different polymer surface models.

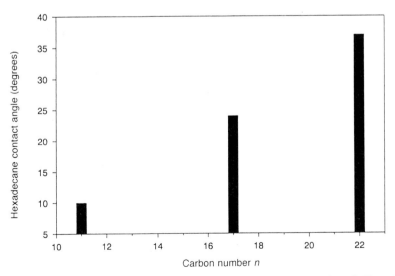

Figure 7. The hexadecane contact angle versus carbon number for $C_nH_{2n+1}O-$ groups grafted at constant coverage to an SiO_2 surface

4 SUMMARY

In this report we have shown how the emerging techniques of molecular self-assembly at surfaces can be utilized to prepare polymer model surfaces. In comparison to bulk polymer surfaces which are difficult to prepare in controlled ways and difficult to characterize quantitatively, these monolayer films can be synthesized with precisely controlled architectures of functional groups and they are ideal for characterization by a large number of surface tools. The systems mentioned have consisted of both polymers and short chain molecules and have consisted of both ordered and disordered assemblies. However, these systems represent only a bare fraction of the structures possible. In this regard the future appears bright for using these self-assembled films to rationally and systematically probe the detailed correlations between molecular structure and properties at polymer surfaces, including even the vastly complex area of biological processes.

REFERENCES

1. For example, see Andrade, J. (ed.), *Surface and Interface Aspects of Biomedical Polymers*, Plenum Press, New York (1985).
2. Briggs, D., and Seah, M., *Practical Surface Analysis by Auger and X-ray Photoelectron Spectroscopy*, John Wiley, New York (1991).
3. Briggs, D., *Br. Polymer J.*, **21**, 3 (1989).
4. vander Wel, H., Lub, J., Van Velzen, P. N. T., and Benninghoven, A., *Mikrochim-Acta*, **II**, 3 (1990).
5. Pireaux, J. J., Vilar, M. R., Brinkhuis, R., and Schouten, A. J., *Langmuir*, **7**, 2433 (1991).
6. Sarid, D., *Scanning Force Microscopy*, Oxford University Press, New York (1991).
7. Bain, C. D., Davies, P. B., Ong, T. H., Ward, R. M., and Brown, M. A., *Langmuir*, **7**, 1563 (1991).
8. Vogel, V., and Shen, Y. R., *Ann. Rev. Mater. Sci.*, **21**, 515 (1991).
9. Jordan-Sweet, J. L., in *Metallization of Polymers*, American Chemical Society Symposium Series 440, Washington DC (1990).
10. Griffiths, P. R., and DeHaseth, J. A., *Fourier Transform Infrared Spectrometry, Chemical Analysis*, Vol. 83, John Wiley, New York (1986).
11. Mackenzie, M. W. (ed.), *Advances in Applied Fourier Transform Infrared Spectroscopy*, John Wiley, New York (1988).
12. Collins, R. W., and Kim, Y. T., *Anal. Chem.*, **62**, 2274 (1990).
13. Schlotter, N. E., in J. I. Kroschwitz (ed.), *Encyclopedia of Polymer Science Engineering*, Vol. 14, John Wiley, New York (1988), pp. 8–45.
14. Gillberg, G. J., *J. Adhes.*, **21**, 129 (1987).
15. Debe, M. K., *Prog. Surf. Sci.*, **24**, 1 (1987).
16. For example, see Ulman, A., *An Introduction to Ultrathin Organic Films, from Langmuir–Blodgett to Self-Assembly*, Academic Press, New York (1991).
17. Laibinis, P. E., Whitesides, G. M., Allara, D. L., Tao, Y. T., Parikh, A. N., and Nuzzo, R. G., *J. Am. Chem. Soc.*, **113**, 32 (1991) and references cited therein.
18. Finklea, H. O., Blackburn, A., Richter, B., Allara, D. L., and Bright, T., *Langmuir*, **2**, 239 (1986).

19. Schlotter, N. E., Porter, M. D., Bright, T. B., and Allara, D. L., *Chem. Phys. Lett.*, **132**, 93 (1986).
20. Nuzzo, R. G., and Allara, D. L., *J. Am. Chem. Soc.*, **105**, 4481 (1983).
21. Nuzzo, R. G., Fusco, F. A., and Allara, D. L., *J. Am. Chem. Soc.*, **109**, 2358 (1987).
22. Nuzzo, R. G., Dubois, L. H., and Allara, D. L., *J. Am. Chem. Soc.*, **112**, 558 (1990).
23. Bain, C. D., Troughton, E. B., Tao, Y. T., Evall, J., Whitesides, G. M., and Nuzzo, R. G., *J. Am. Chem. Soc.*, **111**, 321 (1989).
24. Allara, D. L., in L. H. Lee (ed.), *Adhesion and Adsorption of Polymers*, Plenum Press, New York (1980), Part B, pp. 751–6.
25. Allara, D. L., and Nuzzo, R. G., *Langmuir*, **1**, 45 (1985).
26. Allara, D. L., and Nuzzo, R. G., *Langmuir*, **1**, 52 (1985).
27. Atre, S. V., and Allara, D. L., manuscript in preparation.
28. Cassie, A. B. D., *Discuss. Faraday Soc.*, **3**, 11 (1948).
29. Strong, L., and Whitesides, G. M., *Langmuir*, **4**, 546 (1988).
30. Chidsey, C. E. D., Liu, G. Y., Rowntree, P., and Scoles, G., *J. Chem. Phys.*, **91**, 4421 (1989).
31. Nuzzo, R. G., Korenic, E. M., and Dubois, L. H., *J. Chem. Phys.*, **93**, 767 (1990).
32. Fenter, P., Eisenberger, P., Li, J., Camillone, N., Bernasek, J., Scoles, G., Ramanarayanan, T. A., and Liang, K. S., *Langmuir*, **7**, 2013 (1991).
33. Allara, D. L., Atre, S. V., Elliger, C. A., and Snyder, R. G., *J. Am. Chem. Soc.*, **113**, 1852 (1991).
34. Atre, S. V., Parikh, A. N., Elliger, C. A., Snyder, R. G., and Allara, D. L., manuscript in preparation.
35. Atre, S. V., and Allara, D. L., unpublished results.
36. For example, see Paynter, R. W., Ratner, B. D., and Thomas, H. R., in S. W. Shalaby, A. S. Hoffman, B. D. Ratner and T. A. Horbett (eds.), *Polymers as Biomaterials*, Plenum Press, New York (1984), p. 121; and also Hearn, M. J., Ratner, B. D., and Briggs, D., *Macromolecules*, **21**, 2950 (1988).
37. Sheen, C. W., Shi, J. X., Martensson, J., Parikh, A. N., and Allara, D. L., *J. Am. Chem. Soc.*, **114**, 1514 (1992).
38. Hair, M. L., *Infrared Spectroscopy in Surface Chemistry*, Marcel Dekker, New York (1967).
42. Parikh, A. N., and Allara, D. L., manuscript in preparation.

3

Non-equilibrium Effects in Polymeric Stabilization

M. E. Cates and J. T. Brooks

Cavendish Laboratory, Cambridge

1 INTRODUCTION

In some respects the effect of adding soluble polymer to colloidal dispersions is now theoretically and experimentally well understood.[1,2] The role of the polymer can be either that of a stabilizer, preserving the dispersion, or that of a flocculating agent, depending upon the solvent and the nature of the adsorption. Homopolymer achieves steric stabilization by adsorbing to particles non-specifically at multiple points, while block copolymers (diblocks or triblocks) adsorb typically at one end only, the rest of the chain remaining in solution.

One may appreciate the nature of the polymeric contribution to the interparticle force by considering the osmotic pressure of the polymer solution in the gap between the surfaces. For irreversibly, terminally anchored chains, compression to intersurface separations less than the thickness of the adsorbed layer produces a local increase in the polymer concentration, and hence an excess osmotic pressure and a repulsive force. With adsorbing homopolymer the situation is more complex, since monomers belonging to the same chain can adsorb on to both surfaces. In full equilibrium the polymer can adsorb/desorb, ensuring that the polymer concentration between the surfaces remains small; attractive bridging forces are then dominant. However, if the homopolymer is irreversibly bound to the surface, as is the case under most experimental conditions, the interaction is dominated by the excess osmotic pressure in the gap, and is repulsive. Various other cases will also be considered below.

Theories to predict these effects quantitatively must address the conformations of the molecules, interactions between monomers and the nature of the equilibrium governing the chains. In this chapter we review some of the main theoretical approaches, discussing terminally anchored chains, telechelic chains

Polymer Surfaces and Interfaces II
Edited by W. J. Feast, H. S. Munro and R. W. Richards
© 1993 John Wiley & Sons Ltd

(anchoring at both ends) and adsorbing homopolymer. We pay particular attention to the nature of the equilibrium describing the chains and its effects on the forces between sterically stabilized particles. Several different theoretical approaches are now available to study these problems; here we mainly focus on the self-consistent field theory (in the long chain, strong-stretching limit) for end-attached chains and to the scaling functional approach for homopolymers.

2 GRAFTED CHAINS

The grafting of flexible neutral polymers on to non-adsorbing interfaces is of both theoretical and experimental interest.[1] For high enough grafting densities the polymer chains stretch away from the surface to avoid overlapping, forming a polymer 'brush'. Polymer brushes play an important role in colloidal stabilization: particles interacting via the van der Waals attraction can be protected against flocculation by grafting a brushy layer on to their surfaces; the brushes of two approaching particles resist overlap, producing a repulsion between the particles.

In the case of colloidal stabilization the interface to which the chains are attached is a solid substrate. This is not the only type of interface to which chains can be grafted; for instance the interface could be between two solvents, between solvent and air or between melts or solutions of homopolymer. However, in this chapter we restrict ourselves to solid substrates, where the chain end may be chemically bonded or terminated by a special chemical group that adsorbs on to the surface.

Theoretical treatments of grafted polymer chains have employed energy balance arguments, analytical and numerical 'self-consistent field' (SCF) methods and computer simulations.[3-7] Energy balance arguments enable macroscopic properties, such as the layer thickness or 'brush height', to be estimated, but can reveal no detailed information about the grafted layer.[3,4] On the other hand, the numerical lattice models, based on the work of Scheutjens and Fleer,[5] provide a very accurate description, results of which are supported by those obtained by Monte Carlo and molecular dynamics simulations.[6,7] The analytical SCF approach was proposed independently by Milner *et al.*[8] and Skvortsov *et al.*,[9] both groups exploiting the fact that the chains in the brushy layer are in a strongly stretched state.[10] Under these conditions of strong stretching, some relatively simple theoretical results can be obtained.

In this section we first discuss the energy balance description of the polymer brush. We then briefly outline the continuous SCF approach of Milner *et al.*, before turning to consider the interaction between brushy layers. Finally, we mention recent experimental and theoretical work on the bridging effects of telechelic brushes.

2.1 Energy Balance Description of the Polymer Brush

A very simple estimate of the balance achieved by the stretched chains, first presented by Alexander[3] and de Gennes,[4] is based on the following argument. Consider a set of flexible polymer chains, with degree of polymerization N, permanently attached at a grafting density of σ chains per unit area to a flat interface and exposed to solvent. If the distance $\sigma^{-1/2}$ between grafting sites is much smaller than the typical chain dimension $R_g = N^{1/2}a$, where a is the monomer size, then the chains will be strongly stretched. The height h of the brush is then 'chosen' to balance the two free energy costs: stretching, which reduces the configurational entropy, and the excluded-volume interaction between monomers.

Within this framework the free energy cost per chain is given by (in units where $k_B = 1$)

$$\Delta f \simeq T\left[\frac{h^2}{2Na^2} + \frac{wNa^6}{2}\left(\frac{N\sigma}{h}\right)\right] \tag{1}$$

where w is the excluded-volume parameter, which measures the strength of the excluded-volume interaction between monomers. (Throughout this chapter, $\simeq$ denotes an equality to within constants of order unity.) The equilibrium brush height h^* and the corresponding free energy per chain f^* are found by minimizing equation (1) with respect to h, giving

$$h^* \simeq Na^2(w\sigma a^2)^{1/3}$$

and

$$f^* \simeq TNa^2(w\sigma a^2)^{2/3} \tag{2}$$

respectively. This is an important result: for a brush at fixed coverage σ the brush height grows linearly with N, while the unstretched chain dimension R_g only grows as $N^{1/2}$. For long enough chains, h^* is much larger than R_g—the chains are strongly stretched.

2.2 The Continuous SCF Approach

As mentioned in Section 2.1, Milner *et al.* used an SCF approach to describe the properties of the grafted brush.[8] Their model is based on a mathematical analogy between the configurations of the grafted chains and the trajectories of particles in a potential field. Furthermore, they exploit the insight provided by the theories of Alexander and de Gennes (see Section 2.1) that the chains are strongly stretched. Under these conditions the fluctuations of a grafted chain about its average configuration become unimportant, allowing the properties of the brush to be determined by the average (or classical) configurations only. The SCF approach relates the potential acting on a monomer to the local structure of its environment. This is achieved via a standard mean-field approximation, valid when the grafting density is sufficiently high and the solvent quality is not too good.

Within this description each grafted chain i in the system may be represented as a set of N monomer coordinates $\mathbf{r}_i(n)$, with $n = 0, 1, \ldots, N$. The chains are considered to follow classical trajectories $\mathbf{r}(\mathbf{u}, n)$, beginning with their free ends at $\mathbf{u}_i = \mathbf{r}_i(0)$, in the potential $U(\mathbf{r}) = -w\phi(\mathbf{r})$, where $\phi(\mathbf{r})$ is the self-consistently calculated monomer volume fraction. The free energy is then determined by substituting the classical trajectories $\mathbf{r}_i(n) = \mathbf{r}(\mathbf{u}_i, n)$ into the following expression:

$$F = T \sum_i \int \left[\frac{1}{2} \left(\frac{\partial \mathbf{r}_i}{\partial n} \right)^2 \right] dn + \frac{wT}{2} \int \phi^2(\mathbf{r}) \, dV \tag{3}$$

The first term is the sum of Gaussian stretching energies, whilst the second accounts for the mean-field excluded-volume interaction between monomers. Finally, the free energy F is minimized over the positions $\mathbf{u}_i$ of the free chain ends to determine the equilibrium state of the brush. The resulting equilibrium profile is surprisingly simple (see Figure 1) and has the following parabolic form:

$$\phi(z) = \phi(0) \left[1 - \left(\frac{z}{h^*} \right)^2 \right] \tag{4}$$

where $\phi(0)$ is the volume fraction at the surface, z is the distance normal to the surface and h^* is the equilibrium brush height, given by

$$h^* = \left(\frac{12}{\pi^2} \right)^{1/3} N a^2 (w \sigma a^2)^{1/3} \tag{5}$$

As can be seen, the brush height h^* scales as $\simeq N \sigma^{1/3}$, in agreement with the energy balance predictions of Alexander and de Gennes (Section 2.1).

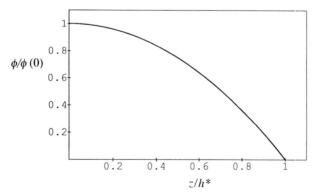

Figure 1. Variation of the monomer volume fraction $\phi/\phi(0)$ with respect to distance from the surface z/h^* for the parabolic brush

Recent experimental work on end-adsorbed brushes, in particular the neutron reflectivity studies of Field *et al.*,[11] indicates that the monomer density is well described by the theoretically predicted parabola. Using a volume fraction profile of the form $\phi(z) = \phi(0)[1 - (z/h^*)^n]$, where $\phi(0)$ and h^* are experimentally determined and n is a fit parameter, they find that good fits to their data (over a range of molecular weights) are obtained for values of n between 1.9 and 2.6. An improvement in the fit can be made by using a two-parameter error function to describe the profile; this is essentially parabolic near the surface but has a decaying tail at large distances. The improved fit is not surprising, since the strong-stretching assumption implicit in the model of Milner *et al.* must fail at large distances from the surface (of the order of the brush height) where the osmotic pressure is small. It should be noted that the parabolic nature of the profile has also been checked by Monte Carlo[6] and molecular dynamics simulations.[7]

2.3 Fixed and Self-adjusting Brushes

Up to now we have discussed the equilibrium structure of an uncompressed brush, making no reference to the nature of the anchoring. In this section we consider the differing properties of irreversibly and reversibly adsorbed chains.

For a chemically anchored brush the grafting density σ is non-adjusting, and is determined by the preparation history of the sample. In contrast, for the end-adsorbed brush the surface coverage could adjust itself to ensure that the system is in its minimum free energy state, but whether the adsorption is truly reversible depends upon the sticking energy of the adsorbing group and also the time scale of the experiment of interest.

Consider the case where end-adsorbed chains in the brush are in equilibrium with 'free' chains in a bulk solution. The thermodynamic potential per unit area Ω/A of the brush is given by[12]

$$\frac{\Omega}{A} = \frac{F}{A}(\sigma) - \sigma(\Delta + \mu_b) \tag{6}$$

where σ is the self-adjusting coverage, μ_b is the bulk chemical potential and Δ is the sticking energy of the anchor group to the surface. Minimization of Ω/A with respect to the coverage, using the form for $F(\sigma)$ given by the model of Milner *et al.*, then yields the equilibrium coverage to be[13]

$$\sigma_{eq} \simeq \frac{1}{wa^2}\left(\frac{\mu'}{Na^2}\right)^{3/2} \tag{7}$$

where $\mu' = (\mu_b + \Delta)/T$. Substitution of the equilibrium coverage into the expression for the brush height (equation 5) gives

$$h^* \simeq a(N\mu')^{1/2} \tag{8}$$

As can be seen, the scaling of $h*$ with degree of polymerization is $h* \simeq N^{1/2}$, in contrast to the irreversibly adsorbed brush where $h* \simeq N$.

Measurements of the dependence of $h*$ upon molecular weight for the end-adsorbing system PEO–PS have been made by Field *et al.*[11] The experimental scaling is observed to be $h* \simeq N^{0.57 \pm 0.06}$, in good agreement with the mean-field prediction (equation 8). It would thus appear that under the time scale of the experimental observations the adsorbing chains are reversibly anchored. This, however, is not true under different experimental conditions, as will be seen below for the force-balance experiments.

2.4 Forces between Grafted Layers

We now consider the force between surfaces carrying grafted brushes, under the differing conditions of irreversible and reversible chain anchoring. We restrict our discussion to planar surfaces (as usual, results for flat geometry can be converted to surfaces with curvature via the Derjaguin approximation).

Consider placing two identical planar surfaces carrying brushy layers a distance $2h$ apart, such that the two brushes are compressed, i.e. $h < h*$. As discussed by Milner *et al.*,[8] there is no interpenetration of the brushes in the classical limit; chains from one brush that penetrate into the other brush pay the penalty of higher stretching and excluded-volume energies due to their encounter with more monomers. Since there is no interpenetration, each of the two brushes can be considered separately, with only the compressed brush height h as a parameter; each brush then has a parabolic density profile, which meets the other at a cusp midway between the grafting surfaces.

In the uncompressed brush (see Section 2.2), every chain has the same free energy. Using this fact, the free energy may be calculated by progressively adding chains until the desired coverage is reached.[8] In a similar fashion, the free energy of the compressed brush is obtained by building up the brushes until they first touch (during which $h(\sigma)$ increases), and then adding the remaining chains (during which h is fixed, while the number of monomers at the surface $\phi(0)$ continues to increase). The resulting free energy per brush, per unit area, within the model of Milner *et al.* is given by

$$\frac{F}{A} = TNa^2\sigma\left(\frac{\pi^2}{12}\right)^{1/3}(w\sigma a^2)^{2/3}\left(\frac{1}{2u} + \frac{u^2}{2} - \frac{u^5}{10}\right) \tag{9}$$

where $u = h/h*$. The force between the two surfaces is given by the osmotic pressure of the polymer in the intersurface gap and can be simply calculated from the free energy expression of equation (9).

2.4.1 Irreversibly Anchored Chains

For irreversibly anchored chains attached to surfaces suspended in a bath of pure solvent, the force per unit area between the surfaces, or disjoining pressure, is given by[8]

$$\Pi_d = -\left.\frac{\partial(F/A)}{\partial(h)}\right|_\sigma \tag{10}$$

Substitution of equation (9) into equation (10) then gives the disjoining pressure for the fixed coverage brush as

$$\Pi_d = T\frac{Na^2\sigma}{h^*}\left(\frac{\pi^2}{12}\right)^{1/3}(w\sigma a^2)^{2/3}\left(\frac{1}{2u^2} - u + \frac{u^4}{2}\right) \tag{11}$$

Under strong compression the first of the three terms in equation (11) is dominant and the disjoining pressure scales as $1/h^2$, as expected, if the excluded-volume interaction dominates over the stretching energy, leading to purely osmotic scaling. Recently Milner[14] has performed a non-adjustable fit of the parabolic brush predictions to the experimental data obtained from the force-balance experiments of Taunton *et al.*[15] The agreement obtained is good, further confirming the validity of the continuous SCF model, applied with the assumption that σ remains fixed during the course of a compression experiment (see below).

2.4.2 Reversibly Anchored Chains

In the reversibly adsorbing case, as was discussed in Section 2.3, the coverage is not fixed, but is governed by the chemical potential of chains in the bulk solution. The net force between the surfaces is given by the excess osmotic pressure in the intersurface gap, i.e. the osmotic pressure in the gap less the osmotic pressure of the bulk. If we assume that the bulk is very dilute, the corresponding osmotic pressure can be neglected and the disjoining pressure between the surfaces is given by

$$\Pi_d = -\left.\frac{\partial(\Omega/A)}{\partial(h)}\right|_{\mu_b} \tag{12}$$

where Ω is the grand free energy and μ_b is the bulk chemical potential, as given by equation (6). In this case the calculated disjoining pressure is smaller than that obtained for the irreversibly anchored brush. Figure 2 shows both the disjoining pressure for the irreversibly and reversibly anchored brush. As can be seen, under strong compression the disjoining pressure for the reversibly anchored brush becomes constant, scaling as $\mu'/2N^2w$;[13] the brush reduces its coverage under compression to maintain a constant volume fraction of monomers in the gap.

Experimental work on end-adsorbing chains using the force-balance technique[12,16,17] provides a probe for the nature of the equilibrium governing the chains. In these experiments the measured force was found to be monotonically

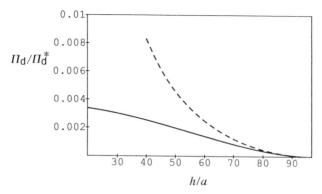

Figure 2. Variation of the disjoining pressure $\Pi_d/\Pi_d^* = \Pi_d a^3/T$ with compressed brush height h/a for irreversibly and reversibly anchored chains. Solid curve: reversible brush; dashed curve: irreversible brush. Values of parameters are: $N = 1000$, $wa^3 = 0.0182$, $\Delta = 30T$, $\phi_b = 10^{-5}$, and coverage at zero compression $\sigma = 0.041$. (Courtesy of D. F. K. Shim[13])

repulsive and far larger in magnitude than the saturated theoretical prediction for the reversible brush. Furthermore, the forces did not vary with the rates of compression/decompression used, indicating that under the time scale of the measurements the chains are essentially irreversibly adsorbed. However, measurements of the equilibrium brush height reveals that, at least in one system, $h^* \simeq N^{1/2}$ (see Section 2.3), indicating that the adsorption is reversible.[12] In contrast, the force law is well fitted by assuming that chains cannot desorb, as mentioned in Section 2.4.1.

This suggests that the time scale for the equilibration of the end-adsorbed chains used by Taunton *et al.*[12,15] (PS–PEO and zwitterion-terminated PS) is long compared to the time scale of a compression experiment (minutes or hours) but short compared to the time scale for which samples are left before measurements are begun (typically overnight or days). Of course, the equilibration rate itself may be much slower when surfaces are in close contact than for a free surface, due to the entanglement of the chains in the gap, and this complicates the discussion.

This slowing-down effect is likely to be very large in the force-balance apparatus where two macroscopic surfaces are used and diffusion of chains over great distances is required. However, it might be much smaller in colloidal systems or copolymer micelles, where the geometry of the interparticle gap is less restrictive, and in these cases one should bear in mind that end-adsorbing chains could be in full equilibrium with chains in a bulk solution. As shown above, this leads to a great reduction in the interparticle force and hence in colloidal stability. It is quite possible that in many systems of interest, the ultimate stability and other static properties, e.g. the structure factor $S(q)$, would then be determined by equilibrium thermodynamics using the reversible brush description for the interparticle forces. On the other hand, dynamical properties of the colloid may depend delicately upon the collision times of the particles and the equilibration times of the adsorbed layers; thus to describe non-equilibrium situations, e.g. flocculation induced by the shearing of a colloidal suspension, the irreversible description of the brushy layers may be more appropriate.

2.5 Telechelic Brushes

For brushes with a single attachment point on each chain, there is always a repulsive interaction between adsorbed layers, at least when the solvent is good. It is interesting to consider also brushy layers for which an attractive component of the force can be devised. This situation arises in the case of telechelic polymers (chains with a functional group at each end) or triblock copolymers of a similar geometry. Such systems have been studied in very recent experimental[18] and theoretical[19,20] work.

Ligoure, Leibler and Rubinstein[20] studied the case of telechelic brushes and assumed that there was a full equilibration between the different possible arrangements of the attachment groups, but that chains could not leave the gap between two surfaces compressed together. The assumed state of an isolated brush was one in which each chain is doubled over to have two attachment points to the surface. In this case the osmotic repulsion between adsorbed brushes is countered by a bridging attraction which arises from chains which have one end adsorbed on to each surface. Within the strong-stretching theory, the energy of a chain that stretches to the midpoint between the two surfaces is the same regardless of whether it (a) continues in a symmetric configuration to the other side of the gap or (b) returns to the original grafting surface in an adsorbed loop. Therefore, a gain in entropy can be obtained by having a certain fraction of bridging chains, even if this means that they are slightly stretched relative to the other components of the brush (see Figure 3). These authors indeed found an attractive region for brushes at a separation comparable to the unperturbed brush height.

In a separate study, Johner and Joanny[19] studied the case of a brush of telechelic chains in which each chain has only one attachment to the surface, the other functional group not being attached. (This could happen if the adsorption occurred in non-equilibrium conditions or if two different types of functional group were present on each chain.) The interaction between such a brush and a bare surface was studied and an attraction was again found at separations of the order of the brush height or less.

Dai and Toprakcioglu[18] studied experimentally the interaction between a triblock copolymer and a bare surface, finding a strong attractive interaction on first bringing the surfaces together. However, if the surfaces were then

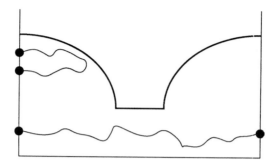

Figure 3. Bridging in telechelic brushes. The bold curve represents the polymer volume fraction between two parallel surfaces; this is parabolic near the surfaces and flat in the central region

separated and brought back together the attraction was reduced, and eventually the interaction became purely repulsive after many compression cycles. This effect cannot be explained by either of the simple models referred to above; it seems to indicate that initially at least some of the adsorbing groups protrude from the brush, whereas finally the repulsive interaction suggests a state in which both surfaces are coated with doubled-over chains. However, to avoid the attraction predicted even for this case,[20] one must assume that final doubling-over of the chain is irreversible. Clearly the experimental situation with telechelic and triblock polymers is a rich one; this represents a promising area for future research.

3 ADSORBED HOMOPOLYMER

The theoretical modelling of homopolymer adsorption has been the subject of much work.[2] Early work considered single polymer molecules near an adsorbing surface, treating the conformations of the molecule as random walks.[21–25] These theories provide a description of the system which includes the polymer–surface interaction and the entropic confinement of the polymer, but do not include the excluded-volume interaction between monomers. Later theories incorporate the excluded-volume interaction via a mean-field approximation[26,27] and facilitate further extension to the many-chain adsorption problem. Several SCF approaches to the problem exist, more recently the continuous theory of Ploehn *et al.*[28,29] and the numerical lattice models of Scheutjens and Fleer.[30–32] These approaches enable detailed properties of the layer to be determined, and can be used to calculate the interaction force between surfaces carrying adsorbed polymer. However, they are somewhat limited, in that despite their complexity they are still only mean-field approximations.

An alternative approach was proposed by de Gennes,[33,34] which is pleasing for its simplicity and provides a convenient framework for discussing a range of equilibrium and non-equilibrium effects. The theory enables the properties of the adsorbed layer to be calculated in terms of a few fundamental parameters, and has been successful in describing many adsorption problems.[35-40] Furthermore, the approach can be formulated either in mean-field or an 'extended' mean-field (scaling) approximation which incorporates the correct scaling properties of semi-dilute solutions. It is well understood that for polymer in a good solvent and under semi-dilute conditions (as is the case in many adsorption problems) mean-field theory is unreliable, and it in fact gives qualitatively wrong answers to several of the problems discussed below. Hence, in what follows we stick to the scaling picture.

In this section we first discuss briefly the adsorption of a single real chain, which serves to define some fundamental concepts. We then turn to the many-chain problem, as described within the scaling approach, and specifically interactions between surfaces carrying adsorbed homopolymer.

3.1 Single-chain Adsorption

We first invoke an energy balance argument[41] to describe the adsorption of a single chain. The free energy of the adsorbed polymer has two competing terms: an entropic confinement term and an energy term resulting from monomers sticking to the adsorbing surface. Let D be the thickness of the adsorbed layer; the entropic confinement is then given by scaling arguments (in units where $k_B = 1$) as $\Delta F \simeq T(R_F/D)^m$ where R_F is the Flory radius of the free chain, which in good solvent is $R_F = aN^{3/5}$ (we have as usual approximated the polymer size exponent $v = 0.588$ by the Flory estimate $3/5$), a is the monomer size and N is the degree of polymerization. The criterion that ΔF is linear in N for extreme confinement ($D \simeq a$) fixes the exponent $m = 5/3$.[42] Turning now to the energy contribution, we write $\Delta F \simeq -T\gamma Nf$, where γ is the sticking energy associated with an adsorbed monomer and f is the fraction of adsorbed monomers. To within an order unity prefactor, we find $f = a/D$, and the free energy of the chain is given by

$$\frac{F}{T} \simeq -\gamma \frac{a}{D} N + \left(\frac{a}{D}\right)^{5/3} N \tag{13}$$

Minimization of equation (13) with respect to D gives the dependence of the adsorbed layer thickness D and free energy F on the sticking energy as

$$D \simeq a\gamma^{-3/2}, \qquad \frac{F}{T} \simeq -N\gamma^{5/2} \tag{14}$$

As would be expected, the size of the adsorbed layer D varies with an inverse power law dependence on the sticking energy γ. The relation $D(\gamma)$ is relevant even for the many-chain problem, as will be seen below, where it defines a fundamental length scale. It should be noted in passing, however, that the exponents predicted above are not quite correct.[43] The fraction of monomers in contact with the surface f is estimated assuming that the concentration profile is regular. Monte Carlo simulations[44] reveal that the profile is weakly singular near the wall, leading to a different dependence of f upon a/D, and thus a different set of exponents than those given in equation (14).[43] This 'proximal' singularity is of no real consequence for the many-chain adsorption problem, and we discuss it no further.

3.2 Many-chain Adsorption—Scaling Approach

For the adsorption of a single chain, the bulk volume fraction of polymer far from the surface is zero. As soon as the bulk volume fraction ϕ_b is non-zero there exists a partitioning of the polymer between the surface and the bulk. The surface coverage, defined as the number of monomers in the adsorbed layer per unit area, increases rapidly with increasing ϕ_b and then reaches a plateau regime, which corresponds to the saturation of the surface[45] (the saturated state occurs for very dilute bulk volume fractions). In this plateau regime the structure of the adsorbed layer is essentially independent of the bulk and can be described by scaling arguments.

For a saturated interface it is well known that the conformations of the adsorbed molecules include large loops that extend to distances of order R_F from the surface[45] (a result of the excluded-volume interaction). These conformations imply a concentration profile that has two regions:

(a) A proximal region ($z \ll D$), z is the distance from the surface and D is the length scale defined in Section 3.1.

(b) A central, semi-dilute region ($D \ll z \ll R_F$) where the profile takes a strongly universal, power-law form.

In the presence of a bulk solution of polymer at non-zero concentration there is a third (outer) region, which we ignore in what follows by assuming that the bulk concentration is very low.

In the central region, the polymer represents a semi-dilute solution, for which the screening length or correlation length can be written as[42]

$$\xi = a\phi^{-3/4} \tag{15}$$

where a is a molecular length scale and ϕ is the volume fraction of polymer. In the central region, a local correlation length $\xi(\phi)$, associated with the local concentration ϕ, can be defined by equation (15). No length scale other than z, the distance from the surface, enters the problem, and thus the only possible scaling of ξ on z is $\xi \simeq z$. This provides an example of a self-similar structure (see Figure 4); the presence of the surface implies that $\phi(z)$ adjusts to a value such that the local screening length $\xi(\phi)$ remains comparable to the distance z

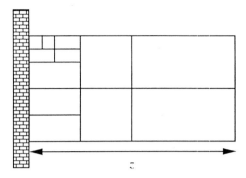

Figure 4. Sketch of the self-similar structure for the adsorbed homopolymer layer. At a distance $z < R_F$ from the surface, the correlation length ξ is of the order of z. (After Ref. 33)

from the surface.[33] Thus the profile is defined by

$$\phi \simeq \left(\frac{a}{z}\right)^{4/3} \tag{16}$$

The self-similar structure of the adsorbed layer, predicted by scaling theory, has now been experimentally verified by Auvray and Cotton[46] using neutron scattering techniques.

The scaling approach outlined above does not allow detailed calculations of (say) the interaction between surfaces carrying polymer layers. To enable precise calculations to be made, de Gennes constructed a functional form for the free energy density of the adsorbed layer, including all the correct scaling exponents, from which definite numerical predictions can be obtained for adsorbed layer properties/interactions. We next discuss the form of the de Gennes functional for adsorption on a single surface from a dilute bulk solution ($\phi_b \simeq 0$); the case of a semi-dilute bulk solution has also been described by de Gennes.[33]

3.3 Adsorption on a Single Surface—Scaling Functional Approach[34]

We define F/A as the free energy per unit area associated with the adsorbed polymer layer. The basic ingredients of F/A are as follows:

(a) A surface contribution describing the direct contact (at $z = 0$) between the polymer and the surface. This contribution can be written as

$$\gamma_d = \gamma_0 - \gamma_1 \phi_s \tag{17}$$

where γ_0 is the surface energy of the pure solvent, γ_1 is the monomer sticking energy ($\gamma_1 > 0$ for adsorption) and ϕ_s is the surface volume fraction of monomers.

(b) An osmotic contribution, related to the free energy density for a uniform solution of volume fraction ϕ. For semi-dilute solutions the interaction free energy density has the structure[42]

$$\frac{F_{osm}(\phi)}{V} = \frac{\beta T}{\xi^3(\phi)} = \frac{\beta T}{a^3} \phi^{9/4} \tag{18}$$

where β is a numerical constant.

(c) A stretching contribution. The fact that the chains do not want to stretch imposes a free energy penalty for gradients in local concentration, which has the following form:[33]

$$\frac{F_{str}(\phi)}{V} = \frac{\alpha T}{\xi^3(\phi)} \left[\frac{\xi(\phi)\nabla\phi}{\phi} \right]^2 = \frac{\alpha T}{\alpha} \phi^{-5/4}(\nabla\phi)^2 \tag{19}$$

where α is another order unity factor.

Adding the contributions (a), (b) and (c), we obtain the structure of the free energy (in units where $k_B = 1$):

$$\frac{F}{A} = \gamma_0 - \gamma_1\phi_s + \frac{\beta T}{a^3} \int_0^\infty [\phi^{-5/4}(m_0 \nabla\phi)^2 + \phi^{9/4}]\, dz \tag{20}$$

where, within the scaling theory, β and $m_0/a = \sqrt{\alpha/\beta}$ are universal constants (independent of chemical microstructure).

Minimization of F/A with respect to changes in the profile $\phi(z)$ results in an Euler–Lagrange equation and boundary condition at the surface. It is interesting to note that from the boundary condition at the surface a length scale, the 'extrapolation length', arises:

$$D = -\left(\frac{1}{\phi} \frac{d\phi}{dz} \right)_s \simeq \left(\frac{T}{\gamma_1 a^2} \right)^{3/2} \tag{21}$$

As can be seen, this length scale coincides with that defined by the energy argument of equation (14). Solution of the profile equations results in a self-similar structure as expected, but now with the volume fraction of monomers in contact with the surface ϕ_s determined by the size of the extrapolation length D; Figure 5 shows a qualitative plot of the volume fraction profile. Finally, substitution of the profile solution into the free energy functional of equation (20) enables the free energy of the adsorbed layer to be calculated.

3.4 Forces between Two Adsorbed Layers

In this section we consider (within the de Gennes approach as outlined above) the forces between surfaces carrying adsorbed homopolymer layers under different equilibrium conditions: these are full equilibrium, restricted equilibrium and doubly restricted equilibrium. As in Section 2.4, we restrict our discussion to planar surfaces.

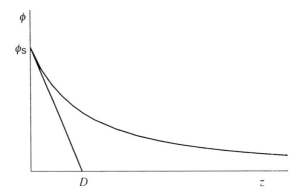

Figure 5. Qualitative plot of the monomer volume fraction ϕ with distance z from a planar surface. The extrapolation length D is governed by the sticking energy of monomers to the surface

Consider placing two identical planar surfaces a distance $2h$ apart in a very dilute polymer solution. By symmetry the free energy functional describing the adsorbed polymer is given by

$$\frac{F}{A} = 2\gamma_0 - 2\gamma_1 \phi_s + 2 \frac{\beta T}{a^3} \int_0^h [\phi^{-5/4}(m_0 \nabla \phi)^2 + \phi^{9/4}]\, dz \tag{22}$$

For calculation purposes it is easier to work with the chemical potential rather than the number of monomers in the intersurface gap, and thus we perform a Legendre transform, introducing the grand free energy per unit area Ω/A:

$$\frac{\Omega}{A} = \frac{F}{A} - \lambda \Gamma; \qquad \Gamma = \frac{1}{a^3} \int_0^{2h} \phi\, dz \tag{23}$$

where λ is the monomer chemical potential and Γ is the surface coverage (the number of monomers per unit area). Substituting the functional form for F/A yields

$$\frac{\Omega}{A} = 2\gamma_0 - 2\gamma_1 \phi_s + 2 \frac{\beta T}{a^3} \int_0^h [\phi^{-5/4}(m_0 \nabla \phi)^2 + \phi^{9/4} - \mu \phi]\, dz \tag{24}$$

where $\mu = \lambda/\beta T$.

As in Section 2.4.2, the net force between the two surfaces is given by the excess osmotic pressure of the polymer in the intersurface gap. Since we are considering a very dilute bulk, the bulk osmotic pressure can be neglected and the disjoining pressure between the two surfaces is given by[34]

$$\Pi_d = -\left.\frac{\partial(F/A)}{\partial(2h)}\right|_\Gamma = -\left.\frac{\partial(\Omega/A)}{\partial(2h)}\right|_\mu \tag{25}$$

which becomes

$$\Pi_d = \frac{\beta T}{a^3} \, \phi_m^{9/4}(h)[\mu(h) - 1] \tag{26}$$

where ϕ_m denotes the volume fraction at the midpoint between the surfaces. Calculation of ϕ_m and μ for a given h and degree of equilibrium now enables the disjoining pressure to be determined.

3.4.1 Full Equilibrium

It was demonstrated by de Gennes[34] that for full equilibrium, where the polymer can reversibly exchange with the external reservoir, the force is attractive for all intersurface separations. This can readily be seen from equation (26); since we are assuming equilibrium with a very dilute bulk ($\mu \to 0$) the disjoining pressure is $\Pi_d = -\beta T \phi_m^{9/4}/a^3$ where $\phi_m > 0$ and hence gives the negative result. This result, however, does not correspond to experimental data where both attraction and repulsion can be found.

3.4.2 Restricted Equilibrium

Since the scenario of full equilibrium seemed experimentally unrealistic de Gennes[34] proposed the restricted equilibrium (RE) model to describe the interacting polymer layers. He argued that under most experimental conditions the polymer is bound to the adsorbing surface by an energy greatly exceeding the thermal energy $k_B T$, which, combined with the general slowness of polymer diffusion, leads to effectively irreversible adsorption. The polymer is then under local equilibrium conditions, except that the amount of adsorbed polymer (surface coverage) is constrained to be a constant. The chemical potential μ and midpoint volume fraction ϕ_m are now governed by the fixed surface coverage constraint. Assuming that the surfaces are saturated leads to a monatomic repulsive force, consistent with experiments in good solvent,[47,48] the calculated asymptotes of the disjoining pressure being given as[34]

$$\Pi_d \simeq \begin{cases} \dfrac{1}{h^3}, & h \gg D \\[2mm] \dfrac{1}{h^{9/4}}, & h \ll D \end{cases} \tag{27}$$

In this limit the effects of the adsorption become negligible, and the system behaves essentially as a slab of polymer solution with uniform concentration.

The RE model was further extended to describe interactions between unsaturated polymer layers by Rossi and Pincus.[37] In this case they found that an attractive minimum develops for large separations, whilst for small separations the force is given, as in the saturated RE model, by the osmotic $1/h^{9/4}$ law. The presence of the attractive minimum results from a bridging attraction, which is most effective when the adsorbed amount is somewhat below saturation. This finding is in good agreement with the numerical work of Scheutjens and Fleer[32] and experimental data which show optimum flocculation for polymer stabilized lattices when the adsorbed amount of polymer is below saturation.[49]

3.4.3 Doubly Restricted Equilibrium

The RE model describes, at least qualitatively, the interaction of adsorbed polymer layers, but its applicability in certain colloidal systems is doubtful. For instance, in a sterically stabilized colloid the polymeric interaction force between particles must depend on the rate of approach between particles, and if this is rapid it is not necessarily appropriate to calculate the force using equilibrium (or RE) thermodynamics. Recent work by the present authors has further extended the de Gennes approach to consider adsorbed polymer layers interacting under conditions in which the adsorbed surface contact density, as well as the surface coverage, is constrained to be fixed.[50] We refer to this as the 'doubly restricted equilibrium' (DRE) model. This provides an example of a *non-equilibrium interaction force* between polymer layers, a force that vanishes when two layers approach very slowly but may be large in the course of a rapid collision, for example in a strongly sheared colloid. The possible importance of such forces has recently been suggested by Lips *et al.*[51]

Several limiting cases can be considered, depending mainly on the equilibration rates for the loops and tails of the adsorbed chains and also that for the surface contacts. The DRE model is appropriate if the equilibration of surface contacts is slow compared to both the collision time between particles and the internal relaxation time (Zimm time) of the adsorbed chains. Although the latter inequality seems unlikely for the adsorbed homopolymer, such a situation may arise if the adsorbed species is a copolymer containing a relatively sparse fraction of functional groups or blocks that adsorb rather strongly to the surface. As shown by Marques and Joanny,[52] the adsorption of random copolymer of this type leads to the same equilibrium concentration profile as for homopolymer adsorption, with a renormalized monomer adsorption energy. To obtain a slow equilibration time for the surface contacts, however, requires that each point be attached to the surface with an energy that is large compared with $k_B T$, and this is likely only in the copolymer case. (Homopolymer adsorption energies are typically a small fraction of $k_B T$ per monomer.) It seems plausible that, with suitable copolymers, the time scale for surface contact equilibration could be large compared to collision times, which, at high shear rates, can be in the order of milliseconds or less.

Application of the de Gennes functional approach yields, as in the RE case, the disjoining pressure between two adsorbed layers. We assume that the surface coverage is saturated, and calculate μ and ϕ_m by applying the constraints of fixed coverage and fixed surface contact density. The calculated asymptotes of the disjoining pressure are found to be[50]

$$\Pi_d \simeq \begin{cases} \dfrac{1}{h^3}, & h \gg D \\[2ex] \dfrac{1}{h^{8/3}}, & h \ll D \end{cases} \tag{28}$$

(The asymptote for $h \ll D$ is calculated using scaling arguments. For this limit the scaling functional is not valid within the DRE model.) The large separation limit is, as would be expected, the same as that found for the RE model. For small separations the predicted force is larger than that of the osmotic RE prediction. In the small h limit the value of the volume fraction at the midpoint must exceed that at the surface by a large factor if the constraint of constant coverage is to be maintained. This is in contrast to the RE model where, for all surface separations, the surface volume fraction exceeds the midpoint value; see Figure 6(a) and (b) for plots of the volume fraction profiles within the RE and DRE models for $h \ll D$. In the DRE model, 'depletion' of the polymer near the surfaces for $h \ll D$ leads to an entropic repulsion, and the system behaves like trapped non-adsorbing polymer, producing the $1/h^{8/3}$ force law.

To understand the possible effects DRE could have we consider a sterically stabilized colloidal dispersion, where the particles are coated with a saturated polymer layer having a slow contact equilibration time. For ordinary Brownian motion we anticipate that the interparticle force will be determined by equilibrium thermodynamics, subject to the proviso of the RE model, and thus do not expect DRE to be relevant in governing Brownian flocculation. However, for a sheared dispersion, where collisions between particles could be rapid compared to the surface contact equilibration time, we might expect the DRE model to be applicable in describing the interparticle force.

Assuming that the particles are spherical and that the 'bare' interaction is van der Waals, we have calculated the interparticle force for sterically stabilized particles, within the RE and DRE descriptions of the adsorbed layer. Figure 7 shows the interparticle force in reduced units as a function of the intersurface separation distance for a Hamaker constant of $A = 20k_BT$. At short enough distances van der Waals forces dominate, giving a deep primary minimum, whilst at larger distances the force goes through a primary maximum. As can be seen, the effect of DRE is to increase dramatically the height of the primary maximum in comparison to the RE case. Increasing the size of the Hamaker constant, and hence the size of the van der Waals contribution, shifts the primary maxima to larger intersurface separations where the RE and DRE

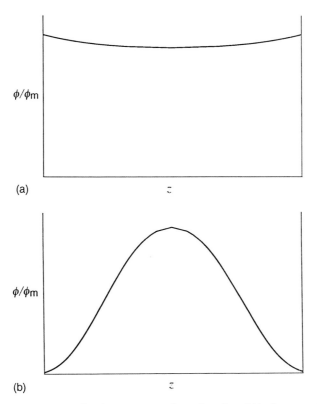

(a)

(b)

Figure 6. Plot of the normalized monomer volume fraction ϕ/ϕ_m between two parallel surfaces (a) within the RE model for $h/D = 0.07$ and (b) within the DRE model for $h/D = 0.08$

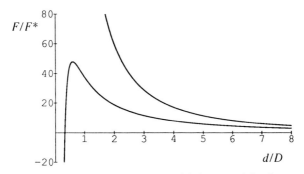

Figure 7. Variation of interparticle force F/F^* with interparticle distance $d/D = 2h/D$ for sterically stabilized particles, calculated using scaling theory for a Hamaker constant of $A = 20T$ ($F^* = \pi R(m_0/a)^3 \beta T/D^2$ where R is the radii of the particles). The upper curve is for the polymer subject to DRE, the lower curve for RE

models coincide. Conversely, as the Hamaker constant is decreased the difference between the RE and DRE force curves becomes more pronounced.

For a dispersion undergoing shear a rough criteria for stability is given by balancing the maximum viscous force with the maximum interparticle force.[53] We would therefore expect that if, for a high enough shear rate, the polymer layer becomes governed by the DRE model, a noticeable increase in the stability of the dispersion to shear will arise.

We emphasize that the non-equilibration of surface contacts is only one of many ways in which non-equilibrium polymer forces could become a determining factor in colloidal stability. However, the model presented above provides an interesting example of such forces which can be calculated quite simply within the existing thermodynamic framework.

4 CONCLUSION

The structure and properties of end-attached and homogeneously adsorbed polymers have been considered within the frameworks of the self-consistent field (with strong-stretching approximation) and the scaling functional theory, respectively. These theories in each case represent a limiting description that applies for extremely long chains. The simplifications brought about in this limit allow quantitative prediction of the force between adsorbed layers under various conditions of equilibration. In the case of end-attached brushes there is a clear distinction between the strong repulsive forces expected at fixed adsorption and much weaker ones that arise when chains are in equilibrium with a reservoir. The latter may prove relevant in discussing the long-time stability of colloids. Bridging forces between telechelic colloids seem experimentally to lead to new non-equilibrium effects. For adsorbed homopolymer in a good solvent, several different types of interaction can be predicted depending on the extent to which the adsorbed layer can equilibrate during an encounter between surfaces. In the thermodynamic model discussed here various limiting cases are accounted for by introducing different statistical ensembles and/or boundary conditions on the interfacial profile. Our recent study of the case where the surface contact density is slow to equilibrate may be relevant to the shear stability of colloids that are stabilized by ionomers or other chains containing functional groups whose dynamics are relatively slow.

ACKNOWLEDGEMENTS

The authors wish to thank C. M. Marques, P. Pincus and D. F. K. Shim for helpful discussions, and also D. F. K. Shim for providing Figure 2. One of the authors (J. T. B.) thanks SERC and Unilever plc for a CASE award.

REFERENCES

1. Milner, S. T., *Science*, **251**, 905 (1991), and references therein.
2. Russel, W. B., Saville, D. A., and Schowalter, W. R., *Colloidal Dispersions*, Cambridge University Press (1989), Ch. 6 and references therein.
3. Alexander, S., *J. Phys. France*, **38**, 983 (1977).
4. de Gennes, P.-G., *Macromolecules*, **13**, 1069 (1980).
5. Cosgrove, T., Heath, T., van Lent, B., Leermakers, F., and Scheutjens, J., *Macromolecules*, **20**, 1692 (1987).
6. Chakrabarti, A., and Toral, R., *Macromolecules*, **23**, 2016 (1990).
7. Murat, M., and Grest, G. S., *Phys. Rev. Lett.*, **63**, 1074 (1989); *Macromolecules*, **22**, 4054 (1989).
8. Milner, S. T., Witten, T. A., and Cates, M. E., *Macromolecules*, **21**, 2610 (1988); *Europhys. Lett.*, **5**, 413 (1988).
9. Skvortsov, A. M., Pavlushkov, I. V., Gorbunov, A. A., Zhulina, E. B., Borisov, O. V., and Priamitsyn, V. A., *Vysokomol. Soedin. A*, **30**, 1615 (1988).
10. Semenov, A. N., *Zh. Eksp. Teor. Fiz. (USSR)*, **88**, 1242 (1985); Eng. Transl.: *Sov. Phys. JETP*, **61**, 733 (1985).
11. Field, J. B., Toprakcioglu, C., Ball, R. C., Stanley, H., Dai, L., Barford, W., Penfold, J., Smith, G., and Hamilton, W., *Macromolecules*, **25**, 434 (1992).
12. Taunton, H. J., Toprakcioglu, C., Fetters, L. J., and Klein, J., *Macromolecules*, **23**, 571 (1990).
13. Shim, D. F. K., unpublished.
14. Milner, S. T., *Europhys. Lett.*, **7**, 695 (1988).
15. Taunton, H. J., Toprakcioglu, C., Fetters, L. J., and Klein, J., *Nature London*, **332**, 712 (1988).
16. Hadziioannou, G., Granick, G., Patel, S., and Tirrel, M., *J. Am. Chem. Soc.*, **108**, 2869 (1986).
17. Ansarifar, A., and Luckham, P. F., *Polymer*, **29**, 329 (1988).
18. Dai, L., Toprakcioglu, C., *Macromolecules*, **25**, 6000 (1992).
19. Johner, A., and Joanny, J.-F., *Europhys. Lett.*, **15**, 265 (1991).
20. Ligoure, C., Leibler, L., and Rubinstein, M., preprint.
21. DiMarzio, E. A., *J. Chem. Phys.*, **42**, 2101 (1965).
22. Rubin, R., *J. Chem. Phys.*, **43**, 2392 (1965).
23. Roe, R. J., *J. Chem. Phys.*, **43**, 1591 (1965).
24. Silberberg, A., *J. Chem. Phys.*, **46**, 1105 (1966).
25. Chan, D., Mitchell, D. J., Ninham, B. W., and White, L. R., *J. Chem. Soc., Faraday Trans. 2*, **71**, 235 (1975).
26. Silberberg, A., *J. Chem. Phys.*, **48**, 2835 (1967).
27. Hoeve, C. A. J., *J. Polym. Sci., Part C*, **30**, 361 (1970); **34**, 1 (1971).
28. Ploehn, H. J., Russel, W. B., and Hall, C. K., *Macromolecules*, **21**, 1075 (1988).
29. Ploehn, H. J., and Russel, W. B., *Macromolecules*, **22**, 266 (1989).
30. Scheutjens, J. M. H. M., and Fleer, G. J., *J. Phys. Chem.*, **83**, 1619 (1979).
31. Scheutjens, J. M. H. M., and Fleer, G. J., *J. Phys. Chem.*, **84**, 178 (1980).
32. Scheutjens, J. M. H. M., and Fleer, G. J., *Macromolecules*, **18**, 1882 (1985).
33. de Gennes, P.-G., *Macromolecules*, **14**, 1637 (1981).
34. de Gennes, P.-G., *Macromolecules*, **15**, 492 (1982).
35. Klein, J., and Pincus, P. A., *Macromolecules*, **15**, 1129 (1982).
36. Ingersent, K., Klein, J., and Pincus, P. A., *Macromolecules*, **19**, 1374 (1986).
37. Rossi, G., and Pincus, P. A., *Macromolecules*, **22**, 276 (1989).
38. Ji, H., and Hone, D., *Macromolecules*, **21**, 2600 (1988).
39. Marques, C. M., and Joanny, J.-F., *Macromolecules*, **22**, 1454 (1989).

40. Brooks, J. T., Marques, C. M., and Cates, M. E., *J. Phys. France II*, **1**, 673 (1991).
41. de Gennes, P.-G., *J. Phys. France*, **37**, 1443 (1976).
42. de Gennes, P.-G., *Scaling Concepts in Polymer Physics*, Cornell University Press, Ithaca, N.Y. (1979).
43. de Gennes, P.-G., and Pincus, P. A., *J. Phys. France*, **44**, L241 (1982).
44. Eisenriegler, E., Kremer, K., and Binder, K., *J. Chem. Phys.*, **77**, 6296 (1982).
45. Bouchaud, E., and Daoud, M., *J. Phys. France*, **48**, 1991 (1987).
46. Auvray, L., and Cotton, J. P., *Macromolecules*, **20**, 202 (1987).
47. Klein, J., and Luckham, P. F., *Macromolecules*, **17**, 1041 (1985).
48. Klein, J., and Luckham, P. F., *Macromolecules*, **18**, 721 (1985).
49. Vincent, B., *Adv. Colloid. Interface Sci.*, **4**, 193 (1974).
50. Brooks, J. T., and Cates, M. E., *Macromolecules*, **25**, 391 (1992).
51. Lips, A., Campbell, I. J., and Pelan, E. G., in E. Dickinson (ed.), *Food Polymers, Gels and Colloids*, Special Publication 82, RSC, London (1991).
52. Marques, C. M., and Joanny, J.-F., *Macromolecules*, **23**, 268 (1990).
53. Russel, W. B., Saville, D. A., and Schowalter, W. R., *Colloidal Dispersions*, Cambridge University Press (1989).

4

Ion Beam Analysis of Composition Profiles near Polymer Surfaces and Interfaces

Richard A. L. Jones

Cavendish Laboratory, Cambridge

1   INTRODUCTION

In conventional surface science it is important that analytical techniques be truly surface sensitive; this is because for metals, semiconductors and small molecule liquids the transition from surface to bulk properties takes place within the first few atomic layers, that is to say within a distance of 5–10 Å from the surface. Traditional surface science techniques such as X-ray photoelectron spectrometry, static secondary ion mass spectrometry, electron energy loss spectrometry and reflection absorption infra-red spectrometry meet this need and are often sensitive only to the first few angstroms of material.[1] In polymer systems, however, for a variety of properties the transition from surface to bulk behaviour takes place over much longer length scales, which are essentially set by the size of the polymer chains; this may be up to a few hundred ångstroms. Moreover, a variety of interesting and technically important phenomena take place at interfaces buried deep within the bulk of the material. Thus, in addition to pure surface analysis techniques, polymer surface and interface science needs techniques that can provide depth profile information in the near surface and interface region. In particular, a variety of phenomena exist whose study requires elemental depth profiling over a range of depths from 10–100 Å to 1 μm from the surface, preferably with good sensitivity and depth resolution. Examples of such phenomena are segregation in polymer blends and mixtures, whether to surfaces, polymer–polymer interfaces or polymer–non-polymer interfaces; surface and interface driven phase separation processes; the kinetics of formation of polymer–polymer interfaces and their intrinsic width; and the kinetics of solvent penetration of polymers.

Polymer Surfaces and Interfaces II
Edited by W. J. Feast, H. S. Munro and R. W. Richards
© 1993 John Wiley & Sons Ltd

2 DEPTH PROFILING TECHNIQUES FOR POLYMER SURFACE AND INTERFACE STUDIES

A number of techniques are now available which go at least part of the way to meeting the need for a high-resolution depth profiling technique. Among them are a closely related group of spectrometries that use as a probe helium ions with energies of about 1–3 MeV; these ion beam techniques are the major subject of this chapter. Before describing them in detail, however, the general requirements for polymer depth profiling will be discussed.

An example of the kind of polymer interface problem that one might be called upon to investigate using a depth profiling technique is illustrated in Figure 1. This is a schematic sketch of a concentration profile through a thin film of a polymer mixture on a substrate—a polymer coating layer on metal, perhaps. There is segregation of one component, both at the free surface, and at the interface with the substrate. What are the instrumental parameters that determine whether, and in what detail, one can detect these effects using a depth profiling technique?

A crucial distinction is between techniques that provide a depth profile in direct space and those which yield information related to the depth profile by some kind of integral transform. In the first category come dynamic secondary ion mass spectrometry and the ion beam analysis techniques, Rutherford backscattering, forward recoil spectrometry and nuclear reaction analysis. In

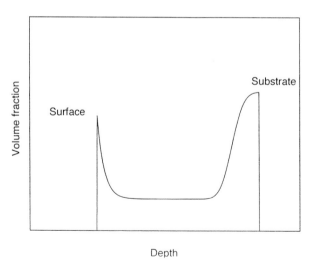

Figure 1. A hypothetical concentration profile of one component of a polymer blend in a thin film deposited on an inorganic substrate, showing segregation both at the surface and at an interface

the second category come neutron and X-ray reflectivity, spectroscopic ellipso-metry, glancing angle X-ray fluorescence and related techniques.

In the case of the direct space techniques the data that emerge from the instrument (in the case of ion beam techniques this is a spectrum of yield versus energy) are related by simple (i.e. essentially arithmetic) operations to the actual depth profile that one is interested in, convoluted with some (typically Gaussian) instrumental resolution function. This is illustrated in Figure 2, which represents the depth profile that might be obtained for the sample shown in Figure 1, convoluted with two Gaussian resolution functions of different characteristic widths (shown in the inset). The characteristic width may be conveniently quoted as the full width at half maximum (FWHM) of the Gaussian, though other definitions are possible, and one must be clear which value is being quoted when making comparisons. For the solid curve the characteristic width is comparable to the length characterizing the segregated layers; at this resolution it is difficult to distinguish the detailed shape of the profiles, though one would be able to extract values of integral quantities such as the integrated surface excess. For the dashed curve the resolution is significantly broader than the length characterizing the segregated layers, and the smaller segregation peak would be difficult to detect at all, particularly if the actual experimental data were at all noisy. Such experimental noise, and any background signal, limit the sensitivity of the technique to small changes in composition. A final factor to be considered is the accessible depth of the technique. Many techniques are only able to analyse within some fixed distance of the surface; therefore, to

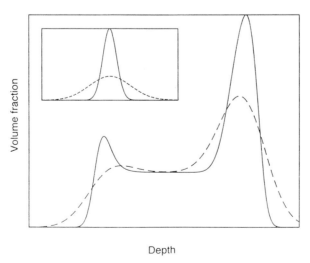

Depth

Figure 2. The concentration profile shown in Figure 1 after convolution with two Gaussian instrumental resolution functions, shown in the inset

detect effects at buried interfaces the substrate–surface distance must be less than this accessible depth.

In contrast, the data from one of the integral techniques is in a form related to the actual depth profile by an integral transform, and simple ideas about resolution are not applicable. In the case of neutron reflectivity,[2] for example, what one measures is the reflectivity $R(k)$ as a function of the wave-vector k; this is the modulus squared of the reflectance $r(k)$, which is related to the scattering length density profile $\rho(z)$ (itself trivially related to the concentration profile) by the following transform:

$$r(k) = -\int_0^\infty \frac{dk(z)}{dz} \frac{1 - r(z)^2}{2k(z)} \exp\left[-2i \int_0^z k(z')\,dz'\right] dz \qquad (1)$$

where

$$k(z) = [k^2 - 4\pi\rho(z)]^{1/2} \qquad (2)$$

This highly non-linear transform is difficult to invert, particularly in view of the fact that the data have an incomplete range of k and phase information has been lost. In practice this type of data is analysed by model fitting. If the experimental conditions have been carefully designed, the data may well be extremely sensitive to the values of the parameters of the model, but there can still be the possibility that a qualitatively different model could fit the data just as well.[3] In the example shown in Figure 1, for example, it might prove difficult to separate the contributions due to the segregation at the surface and at the interface with the substrate. This possibility can be guarded against by carrying out complementary experiments with techniques such as ion beam analysis, with lower ultimate resolution, but with simple data analysis yielding direct space concentration profiles. Another problem with techniques based on reflectivity measurements is that, while they are very sensitive to the details of sharp composition gradients, they are much less sensitive to more gradual changes in composition. In this they are entirely complementary to direct space techniques.

3   ION BEAM ANALYSIS TECHNIQUES

3.1   Introduction

In this section we discuss the ion beam techniques in more detail. These techniques are well established in the materials science community, so here only the general physical principles will be outlined, followed by some comments about practical requirements of instrumentation, with particular reference to the special problems posed by polymers.

3.2 Rutherford Backscattering

The simplest of the techniques, Rutherford backscattering (RBS), provides an excellent way of depth profiling relatively heavy elements in a matrix of lighter elements. As most polymer problems do not tend to involve heavy atoms the technique has not been very widely used for polymers. Despite this, we discuss it first, as it illustrates the principles common to all the ion beam techniques in a straightforward way. Fuller details may be found in the authoritative book by Chu, Mayer and Nicolet.[4]

The basis of RBS is the observation that the interactions between matter and charged ions of energies of a few million electronvolts (eV) fall into two distinct categories. The first type of interaction consists of collisions between ions and the electrons in the sample; these collisions are very frequent and provide the predominant mechanism for the continuous and smooth loss of energy that an ion undergoes on traversing a material, leading to complete stoppage after some well-defined distance. The second type of interaction is the direct collision of the ion with a nucleus in the material. Because of the localized nature of the nucleus such collisions, though rare, lead to the elastic scattering of the ion through large angles (Rutherford scattering).

These rare, nuclear collisions are elastic (and, at the energies we are interested in, non-relativistic) and their kinematics can be calculated using classical considerations of conservation of energy and momentum. It is easily shown that the energy of an incident alpha particle after collision with a nucleus in the sample is a fraction of the initial energy, the value of that fraction being a function of the masses of the incident and target particles (m_1 and m_2 respectively) and the scattering angle θ. This fraction is known as the kinematic factor; for $\theta = 180$ it is given by the expression

$$K = \left(\frac{m_2 - m_1}{m_2 + m_1}\right)^2 \tag{3}$$

The dependence of this factor on target nuclear mass gives the technique its analytical capability—ions are backscattered from heavier ions with a greater proportion of their initial energy than from lighter ions. The energy spectrum of ions backscattered from a thin foil would thus consist of a series of discrete peaks, each peak corresponding to a different nuclear species present in the foil.

For a sample of finite thickness we must also consider the effect of electronic collisions. Only very small proportions of the incident ions undergo nuclear scattering right at the surface of the sample; most penetrate into the bulk of the material, losing energy continuously by electronic collisions on the way. At any depth a proportion of ions will undergo nuclear collisions and be backscattered; the backscattered ions will then lose further amounts of energy before re-emerging from the sample. The differential energy losses with path length for ions going through various materials, due mainly to electronic interactions, are known as 'stopping powers'; these are material parameters which are tabulated as a function of energy for each element. With knowledge of these stopping powers the depth at which a scattering event occurred may be deduced from the measured energy. This is the basis for depth profiling capability of ion beam scattering techniques.

Finally, we must consider what proportion of incident ions are backscattered. If the interaction between the incident ion and the nucleus is entirely due to the Coulomb repulsion due to their electrical charges, the scattering cross-section (the proportion of incident ions scattered into a unit solid angle at a given scattering angle) is given by the Rutherford formula. The assumption of a pure Coulomb interaction relies on the neglect of the screening effects of the electrons, which is permissible for large-angle scattering events at high energies, and on the neglect of the effects of short-range nuclear forces, which is permissible for heavy target nuclei at not too high incident energies. As it happens the conditions typically used for Rutherford backscattering always satisfy the first condition and very often satisfy the second; in these cases the Rutherford formula accurately predicts the cross-sections. Cases in which nuclear forces are important are discussed below, when we come to the technique of nuclear reaction analysis.

The cross-section given by the Rutherford formula has the following form:

$$\text{Cross-section} = \left(\frac{Z_1 Z_2 e^2}{16 \pi \varepsilon E} \right)^2 f(\theta) \tag{4}$$

where Z_1 and Z_2 are the atomic numbers of the incident and target ion, E is the incident energy and $f(\theta)$ is a strong function of the scattering angle, which depends on θ approximately as $\sin^{-4}(\theta/2)$. (Strictly it also contains a weak dependence on the mass ratio as well.) The dependence on the atomic number of the target ion shows that Rutherford backscattering is most sensitive to heavy ions.

Although the principles of RBS are very simple, actual calculations of the spectra expected for a given sample can be lengthy, because the stopping power is a function both of composition and of energy. However, the calculations are easily automated and powerful analysis programs are widely available (e.g. RUMP,[5,6] available from the Materials Science Department, Cornell University).

Without doing detailed calculations, let us consider the factors that determine the depth resolution of the technique. What one measures is the energy of the scattered ions, so the starting point is the energy resolution of the detector and associated electronics, and the energy spread of the incident beam. Additional effective degradation of the energy resolution comes from a geometrical factor; due to the divergence of the beam and the finite acceptance angle of the detector, there is a spread of effective scattering angles. The relation of this spread of angles to an effective spread of energy depends on the derivative of the kinematic factor with respect to angle. Finally, to convert the total effective energy resolution to a depth resolution, one needs to use the stopping power of the material in question. Another factor comes into play for analysis of buried interfaces; electronic energy loss is a statistical process, the sum of a large number of small events, so there is an associated energy spread around the mean energy loss which increases for increasing path lengths. This is called straggling. The overall resolution predicted by this kind of calculation depends strongly on the material being analysed. As an order of magnitude, a typical depth resolution achieved for the analysis of low concentrations of heavy atoms in a hydrocarbon matrix might be around 200 Å FWHM.

To illustrate these points, Figure 3(a) shows an RBS spectrum from a photoresist polymer that has been exposed to the solvent trichloroethane (the physical significance of this spectrum will be discussed below). The concentration profile of trichloroethane that can be deduced from this spectrum is shown in Figure 3(b). The major features of the spectrum are the steep increases of yield with decreasing energy. These represent the energies at which helium ions are backscattered from the surface of the sample from chlorine, oxygen and carbon, respectively. The spectrum between the chlorine edge and the oxygen edge represents backscattering from chlorine atoms at various depths from the surface; the close relationship between the spectrum and the deduced concentration profile is obvious. Note that what in reality is a sharp rise in chlorine concentration at the surface of the sample is broadened in the energy spectrum by the finite energy, and thus depth, resolution of the experiment. The accessible depth in this case is determined by the energy difference between the chlorine edge and the oxygen edge; ions scattered from chlorine atoms deep within the sample will have the same energies as ions scattered from oxygen atoms closer to the surface.

3.3 Forward Recoil Spectrometry

The usefulness of RBS is limited by its lack of sensitivity to light elements. One type of elemental contrast that is particularly useful in polymer physics is provided by deuterium labelled polymers, which may be used as a non-perturbative label, or in studies of isotope effects. Forward recoil spectrometry (FRES), also known as elastic recoil detection analysis (ERDA), is an ion beam analysis technique specifically for depth profiling deuterium and hydrogen.[8–10]

Kinematics makes it impossible for helium ions to be scattered through an angle of more than 45° after a collision with deuterium. The separation between elements would be prohibitively small if one attempted to measure helium ions scattered at an angle of less than 45°. However, one can also consider detecting, rather than the scattered helium ions, the deuterium and hydrogen ions recoiling from collisions; the energy at which a particle of mass m_2 recoils at a scattering

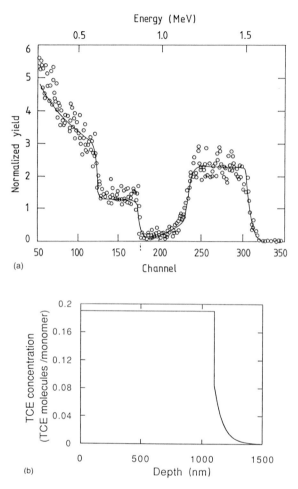

(a)

(b)

Figure 3. (a) Rutherford backscattering spectrum for a photoresist polymer exposed to the solvent trichloroethane for 800 seconds at 20.5°C. The solid line is a simulation corresponding to the solvent concentration profile shown in (b). (Adapted from Mills *et al.*[7] Reproduced by permission of Chapman and Hall Ltd)

angle θ after being hit by a particle of mass m_1 moving with energy E_0 is

$$K_r E_0 = \left[\frac{4m_1 m_2}{(m_1 + m_2)^2} \right] \cos^2(\theta) E_0 \tag{5}$$

FRES is typically carried out with a scattering angle of 30°, with the beam incident on the sample at a glancing angle of 15°. In this geometry equation (5) gives the values $0.5E_0$ and $0.67E_0$ for the recoil energies for hydrogen and deuterium, respectively. One is left with the difficulty of separating the contribution of the forward scattered helium ions from that of the hydrogen and deuterium ions; the surface barrier detectors used for particle detection discriminate only by energy and not by mass. This is a serious problem as the cross-section for forward scattering of helium ions is very much greater than that for recoil of deuterium and hydrogen.

The simplest solution to separating out these contributions lies in the use of a stopper foil—a film of material placed in front of the detector of sufficient thickness to completely stop helium ions while allowing the more penetrating hydrogen and deuterium ions to pass through, albeit with some loss of energy. A very commonly used choice is mylar foil, which has the advantage that it is readily available in uniform, pinhole free sheets. A 10 μm thick sheet is sufficient to stop 3 MeV helium ions while allowing hydrogen and deuterium through. The drawback of using a stopper foil is that the resolution is degraded by energy straggling in the foil. Typically these lead to an effective energy resolution of 50 keV, rather than the 20 keV that can be obtained in RBS without undue effort. This energy resolution translates to a depth resolution of about 800 Å in polystyrene.

Equation (5) shows that the kinematic factor is very sensitive to angle. This does not lead to undue geometrical degradation of the depth resolution for a reasonable detector solid angle, simply because the resolution is dominated by the effect of straggling in the film. However, one must still be aware that small variations in the position of the sample or the beam can lead to appreciable changes in the kinematic factors.

3.4   Time-of-flight Forward Recoil Spectrometry

We have seen that the limit on resolution for the forward recoil technique is set by straggling in the mylar film; this is a fundamental limit in the sense that the straggling introduced by the film is directly related to the total stopping power of the foil, so that the straggling in the foil cannot be reduced without making the foil too thin to stop the helium ions. Improvement in resolution can only come by finding another way to separate the helium ions from the protons and deuterons. Magnetic separation is a possibility, but it turns out that the required fields are rather higher than are convenient. However, as the masses of the particles are different, if one can measure simultaneously the speed and energy of each particle one can discriminate between different masses. Figure 4 shows a sketch of the detector arrangement and electronics required to do this.[11–13] The start signal for the time of flight of a particle is provided by the secondary electrons emitted from a thin carbon foil as the particle passes through (this foil is much thinner than the stopper foil used in conventional FRES, and degradation of resolution due to straggling within it is negligible), which are accelerated and detected by a channel plate electron multiplier. The

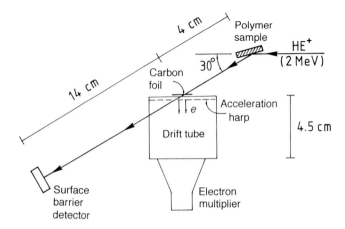

Figure 4. Detector arrangement for the time-of-flight forward recoil technique. (Reproduced from Sokolov *et al.*[11] by permission of the American Institute of Physics)

stop signal is provided by a fast timing pulse from the surface barrier detector, which also provides the energy of the particle. A time-to-analogue converter, activated by these start and stop signals, provides a pulse whose voltage is proportional to the time of flight, and thus inversely proportional to velocity. It would be possible to collect a two-dimensional spectrum, recording yield as a function of both energy and velocity, allowing a complete separation of hydrogen, deuterium and helium ions. It is easier simply to gate out the slower helium ions electronically, however, leaving a spectrum composed only of hydrogen and deuterium ions. The resolution obtainable in this way is determined by the energy resolution of the detector and electronics and by the geometrical factors, arising mainly from the finite acceptance angle of the detector. A resolution of 250 Å FWHM has been obtained for polystyrene films. One drawback of this technique as implemented so far is a rather high background arising from false coincidences, which reduces the effective sensitivity of the technique, though it is believed that this problem is not fundamental and will be reduced by further design improvements.

3.5 Nuclear Reaction Analysis

At higher energies, or for collisions between light ions, the so-called 'Coulomb barrier' may be penetrated, nuclear forces come into play and scattering cross-sections become non-Rutherford. This is exploited in a number of variants on the ion beam scattering technique collectively known as nuclear reaction analysis (NRA).

   The simplest exploitation of non-Rutherford effects lies in the use of nuclear resonances to enhance the sensitivity of Rutherford backscattering and forward recoil spectrometry. The cross-section for helium scattering from deuterium is much greater than predicted by the Rutherford formula, for example, and has a strong maximum, or resonance, at a little less than 2.2 MeV for a scattering angle of 30°.[4] This resonance can be exploited to enhance the sensitivity of the FRES technique to deuterium by tuning the beam energy so that at the depth of interest the cross-section for deuterium is maximized. Similarly, there is a resonance for helium on oxygen at about 3.05 MeV; this has been used to good effect to enhance greatly the sensitivity of RBS to oxygen, which as a relatively low atomic number material otherwise can be difficult to detect using RBS. Particularly at energies greater than 3 MeV, there is a large number of resonances which may be exploited in a similar way.

In addition to this type of resonance, in which the cross-section is enhanced at the resonant energy but in which the collision is still elastic, one can also have resonant nuclear reactions in which energy is released. For example, a ^{3}He ion incident on a deuteron can undergo a nuclear fusion reaction, producing a ^{4}He ion, a proton and a substantial release of energy. The reaction may be written as

$$^3\text{He} + {}^2\text{H} \longrightarrow {}^1\text{H} + {}^4\text{He}, \qquad Q = 18.352 \text{ MeV}$$

and the kinematics of the collision calculated in a similar way as for elastic collisions, by conserving momentum and energy, taking into account the energy Q released in the reaction. Figure 5 shows the results of such a calculation: Figure 5(a) shows the energy of the proton emitted at an angle of 165° as a function of ^{3}He energy, while Figure 5(b) shows the energy of the ^{4}He ion emitted at an angle of 30° as a function of ^{3}He energy. If a beam of ^{3}He ions is incident on a thin film containing deuterium, the emitted spectrum of either ^{4}He or protons gives one depth profile information; as the ions penetrate deeper into the sample they lose energy by electronic collisions, as in RBS or FRES, and if they subsequently undergo a nuclear reaction the particle is emitted at the energy appropriate to the energy of the incident particle, which is simply related to the path length of material it has traversed. The calculations are thus very similar to those for FRES and RBS. Let us first consider the case where one detects protons at 165°; this was the geometry used in the original description of this technique[15] and which is used for polymer work by the Surrey group.[16] As the ^{3}He energy decreases the energy of the emitted proton increases; thus protons emerging from collisions deeper in the material have higher energies than those resulting from collisions on the surface.

In contrast, if one detects ^{4}He at 30°, as done by the Weizmann group,[17,18] particles from deeper collisions emerge with lower energies. An interesting feature of the kinematics in this case is that a given energy loss of the incident ^{3}He ion produces a larger energy loss in the detected ^{4}He ion; this so-called

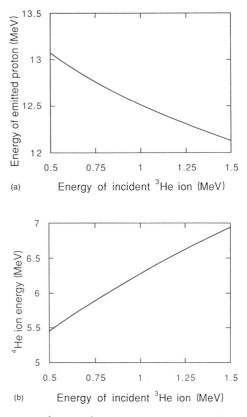

Figure 5. Kinematics of the ^{3}He(d,p)^{4}He reaction. (a) Energy of protons emitted at 165° as a function of energy of an incident ^{3}He beam. (b) Energy of the ^{4}He ions emitted at 30° as a function of energy of an incident ^{3}He beam

'energy amplification' effect leads to an increase in the potential depth resolution. On the other hand, the variation of ^{4}He energy with angle at 30° is greater than the variation of proton energy with angle at 165°; thus geometric contribution to the resolution is more important for the Weizmann geometry than for the Surrey geometry. Figure 6 shows the results of an illustrative calculation showing the depth resolution as a function of detector acceptance angle for both geometries. What this diagram shows is that the highest ultimate resolution is obtainable in the Weizmann geometry, whereas if one increases the detector acceptance angle to improve the count rate, and thus the sensitivity of the technique, the Surrey geometry has the advantage (the reaction cross-section is virtually independent of angle). The best reported resolution for the depth resolution at the surface of a polystyrene film is 140 Å FWHM for the Weizmann geometry.[17] For the Surrey geometry a resolution of 300 Å FWHM

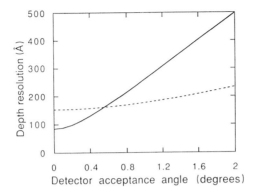

Figure 6. A simple illustrative calculation of the effect of geometrical factors on the depth resolution of the ^{3}He reaction analysis technique. The depth resolution at the surface of polystyrene is plotted as a function of the acceptance angle of the detector (a) for detection of protons emitted at 165°, assuming resolution of the detector and electronics of x keV, and (b) for detection of ^{3}He ions at 30°, assuming resolution of the detector and electronics of x keV. In both cases it is assumed that the sample is tilted at a 15° glancing angle to the incident beam

has been obtained.[16] An advantage of the Surrey geometry over the Weizmann geometry is the much greater accessible depth, which (at lower resolution) can be as great as 8 μm.

3.6 Instrumentation

All the techniques require one to accelerate helium ions to an energy of up to 3 MeV or so and to detect ions scattered from or recoiling out of the polymer sample. The instrumentation required is relatively simple and may be used for all the techniques.

A helium beam of a few MeV in energy may be supplied by one of a variety of small accelerators. Much early ion beam analysis work was carried out on Van der Graaf accelerators that were originally built for nuclear work. Some of these veteran instruments are more than forty years old yet are still working reliably. However, in recent years a new generation of compact accelerators, purpose built for analysis work, have come on to the market, and in fact it is now possible to buy a complete turnkey ion analysis instrument for a comparable sum of money to that required for an advanced analytical electron microscope.

The helium ion beam for such an instrument, based on a modern compact tandem accelerator, is supplied from a radio frequency source and passed over a charge exchange canal to provide singly charged negative ions. These are accelerated up to the terminal voltage, which might be up to one million volts. At the terminal the electrons are stripped from the ions, leaving at least some of them doubly positively charged, and are accelerated away from the positively charged terminal to acquire their final energy, which is, in electron-volts, three times the value of the terminal voltage. The resulting ion beam is then bent by a magnetic field and collimated to provide a well-defined beam of ions of a single mass and charge state.

The sample chamber should ideally be provided with a load lock, enabling the sample to be changed without breaking the vacuum in the chamber. The sample holder should be mounted on a goniometer, and may be cooled in order to reduce beam damage to polymeric samples. A continuous flow of liquid nitrogen is usually sufficient to reduce beam damage to acceptable levels in moderately susceptible materials, though some materials (particularly halogenated polymers and polymethyl methacrylate) suffer mass loss even at these temperatures. Closed-loop helium refrigerators have been used to obtain even lower temperatures. In some cases it is useful to have cooling in order to immobilize the sample and to stop diffusion in polymer–solvent systems and in polymer systems with low glass transition temperatures. In these cases it is essential to have a load lock in order to transfer the sample without its temperature rising to room temperature. The presence of a cold surface in the vacuum chamber presents limitations on the quality of vacuum required. To perform ion beam analysis there is no need in principle for the vacuum to be any better than 10^{-5} torr (the upper limit being set in the chamber by the need to avoid glow discharge around the surface barrier detector and in the accelerator by consideration of the lifetime of the high voltage tubes). However, if the vacuum is poor, extensive build-up of condensed vapour on the sample will take place. This may be avoided either by having a much better vacuum in the chamber or, more straightforwardly, by improving the vacuum locally near the sample holder using a cold shield.

For an accurate absolute quantification of concentration, one must be able to make an accurate measurement of the total charge accumulated on the sample. The sample holder should be electrically insulated from the chamber and connected to a sensitive current digitizer and counter; the sample holder should be surrounded by a suppression cage at a potential of about -400 V so that the charge integration is not rendered inaccurate by the emission of secondary electrons from the sample. One can also sometimes use an internal calibration of total dose, e.g. by measuring an RBS spectrum at the same time as an FRES spectrum.

Particle detectors are usually surface barrier silicon detectors or ion-implanted detectors. These typically have an energy resolution of between 12 and 20 keV, when connected to standard nuclear electronics. The best resolution is usually obtained by isolating the detector earth from the chamber to reduce the possibility of ground loops, which may often be troublesome. Data are collected using multichannel analysers; it is convenient now to use PC-based MCAs. This makes the transfer of data to the data analysis computer straightforward.

4 APPLICATIONS OF ION BEAM ANALYSIS TO POLYMER SURFACE AND INTERFACE PROBLEMS

Although ion beam analysis techniques have been in widespread use in materials science since the 1960s, it is only in the last ten years that their utility for the study of polymer surfaces and interfaces has been realized. At first problems of polymer diffusion received the most attention, but in the last few years a wider variety of polymer surface and interface problems have been attacked using these techniques. The major areas of application of ion beam techniques to polymer surface and interface problems can be divided into three broad classes.

4.1 Segregation at Polymer Surfaces and Interfaces

In polymer mixtures, as in any other type of mixture, the composition at a surface or interface is not in general the same as the bulk composition; if one component of the mixture has a lower surface or interface energy than the other components the system can lower its overall free energy by having a composition at the surface or interface richer in the lower surface or interface component than the bulk. Perhaps the simplest situation arises in a binary, miscible polymer blend in which one component has a lower surface energy than the other; even though the blend is miscible one expects the lower surface energy material to segregate to the surface. Key parameters of interest (see Figure 7) are the actual surface composition, the total adsorbed amount z^* and the shape of the near-surface depth profile. One example of this phenomenon was first detected using forward recoil spectrometry;[19] mixtures of high molecular weight deuterated and protonated (normal) polystyrenes were found on annealing to have

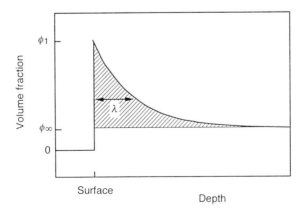

Figure 7. Idealized surface segregation profile, illustrating the surface composition ϕ_1, the bulk composition ϕ_∞ and the decay length λ. The shaded area is the surface excess, z^*

an excess of the deuterated material at the surface (Figure 8). This is the result of a very slight isotopic difference in surface energies between the two species, which arises from the same polarizability differences that lead to a slight unfavourable thermodynamic interaction between the species,[10] leading to phase separation at high molecular weights (the molecular weights used in this particular experiment were not high enough for the system to phase separately in the bulk). Forward recoil spectrometry experiments allow one to measure the total adsorbed amount of d-PS, z^*, as a function of the bulk volume fraction. From this adsorption isotherm it was possible, using a mean field theory of surface segregation,[21] to extract a value for the difference in surface energy between the two components; the value, 0.08 mJ/m^2, was of a similar order of

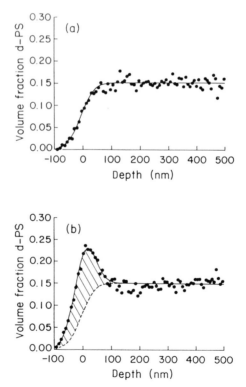

Figure 8. Depth profiles of d-PS in a blend with PS with an initial volume fraction of 0.15, as revealed by forward recoil spectrometry. (a) As spun; the solid line is the fit to a uniform 0.15 blend. (b) After annealing for 5 days at 184°C. The solid line is a fit assuming surface enrichment of d-PS; the surface excess is varied to produce the best fit. Enrichment depth profiles of different shapes, but with the same surface excess, cannot be distinguished. (Reproduced from Jones *et al.*[19] by permission of the American Physical Society)

magnitude to measured values in surface energy between low molecular weight hydrocarbons and their deuterated analogues.[22] However, the depth resolution of the simple forward recoil experiment, at 800 Å, is not good enough to compare the experimental depth profile with the prediction of theory, which is that the depth profile should be approximated exponential with a decay length of the same order as the bulk correlation length for concentration fluctuations in the blend, here about 200 Å. The use of time-of-flight forward recoil spectrometry,[11,13] with its resolution of 250 Å, does allow one to distinguish more details of the shape of the profile, particularly for slightly larger bulk volume fractions for which the correlation length is somewhat longer (Figure 9). Better resolution is obtained using dynamic SIMS[13,23] with a resolution that can be as little as 80 Å, but this improvement in depth resolution is obtained at the expense of slightly less straightforward quantification. Ultimate resolution is obtained using neutron reflectivity;[24] here one is able to see slight deviations in the detailed shape of the profile from that predicted by mean field theory. However, confidence in the results of any one of these techniques is enormously increased by agreement between the different techniques.

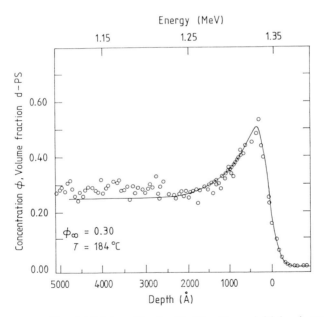

Figure 9. Depth profile of d-PS in a blend with PS with an initial volume fraction of 0.30 after annealing for 5 days at 184°C, as revealed by time-of-flight forward recoil spectrometry. The solid line represents a depth profile decaying exponentially from a surface composition of 0.76 to a bulk composition of 0.26, with a decay length of 600 Å. The depth resolution is 240 Å FWHM. (Reproduced from Sokolov *et al.*[11] by permission of the American Institute of Physics)

In practical situations one is concerned as much with the kinetics by which material segregates to the surface of a blend as with the equilibrium amount; in practice the time spent in a processing step would rarely be long enough for full equilibrium to be reached in a slow-moving polymer system. Forward recoil spectrometry has also been used to study the kinetics of surface segregation in the d-PS/h-PS system[25] and has shown that the kinetics of build-up is limited by diffusion of d-PS from the bulk. Quantitative agreement with theory is obtained.

A number of lessons may be drawn from this extensive series of studies on what on the face of it is a polymer blend system of only academic importance. Firstly, the size of the segregation effects obtained due to a tiny difference in surface energies strongly underlines the importance of surface segregation effects in mixed polymer systems. The amount of material segregated is determined by a balance between the free energy saved by having the lower surface energy material on the surface and the free energy cost of maintaining a surface layer of different composition to the bulk. Because the free energies of mixing of polymer blends are in general small due to their small entropies of mixing, this energy cost is small and may be overcome even by quite small surface energy differences. Extensive studies on the very simple model system d-PS/h-PS, for which the bulk thermodynamics of mixing is better understood than for more complicated blends of chemically different polymers, have allowed detailed and quantitative tests of theory to be made. Finally, these studies have shown clearly how different depth profiling techniques can be used together to exploit their complementary strengths.

Another example of the way in which the ease of analysis of a direct space depth profiling technique such as forward recoil spectrometry can complement neutron reflectivity is provided by studies of segregation of block copolymers to a polymer-inorganic interface. The grafting of polymers by one end or at various points along a chain to a polymer–inorganic interface will have profound implications for adhesion at such interfaces. What is needed is information about the total amount of anchored chains, the shape of the grafted layer, and particularly the sharpness of the interface between grafted and free chains, and the kinetics of the arrival of such chains. Model systems can be constructed by labelling with deuterium chains with 'sticky ends', i.e. chemical groups or short blocks with an affinity for the inorganic material. In one such recent study d-PS chains with short polybutadiene blocks on one end were blended with normally terminated protonated (normal) polystyrene and spun as thin films on to silicon substrates.[26] After annealing, FRES measurements clearly showed segregation of material to the interface between the polymer and the silicon. Once again, the resolution of FRES was not sufficient to determine anything about the detailed shape of the composition profile; it allowed only a measurement of the total amount of segregated material. Neutron reflectivity measurements, however, allowed measurements of the composition profile near the silicon interface to be made in sufficient detail to make a quantitative test of theory. Here the particular importance of doing forward recoil spectrometry measurements as well as neutron reflectivity measurements is that the reflectivity measurements cannot by themselves distinguish between adsorption at the surface and adsorption at the interface.[3] Comparison between the adsorbed amounts as measured by FRES and NR gives a further check on consistency.

Segregation of block copolymers at polymer–polymer interfaces is another area in which measurements made by forward recoil spectrometry can give valuable information. Block copolymers have long been recognized as 'compatibilizers' of immiscible polymer mixtures; by segregating at the interface between the immiscible polymers they lower the interfacial energy and in some circumstances increase the fracture toughness of the interface.[27,28] Scaling and mean field theories predict the segregated amount of block copolymers and their conformation at the interface.[29] By preparing a planar interface between two incompatible polymers, one of which contains dispersed within it deuterium-labelled block copolymer, and annealing the resulting sandwich, adsorption of block copolymers at the polymer–polymer interface can be detected and the adsorption isotherm can be measured. Figure 10 shows the results for a deuterium-labelled styrene–vinyl pyridine block copolymer at an interface

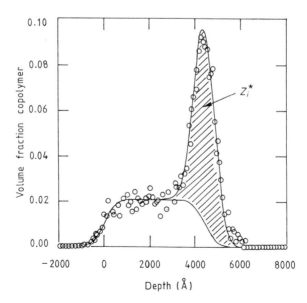

Figure 10. Concentration profile of d-PS–PVP block copolymer in a sandwich of PS and PVP, as revealed by forward recoil spectrometry. (Reprinted with permission from Shull *et al.*[30] Copyright (1990) American Chemical Society)

between polystyrene and poly(vinyl pyridine).[30] Figure 11 shows the adsorption isotherm; it has the familiar shape expected both for block copolymers and for surfactants, becoming markedly less steep at the critical micelle composition. Samples with bulk concentrations of block copolymer greater than the critical micelle composition show segregation of the block copolymer both to the polymer–polymer interface and to the surface. Surface segregation appears only for bulk compositions higher than the critical micelle composition; the homopolymer forming the topmost layer here is polystyrene, which has a lower surface energy than poly(vinyl pyridine). Together these facts suggest that segregation of whole micelles takes place at the surface; this has now been confirmed by electron microscopy.[31]

Moving back now to simple homopolymer mixtures, we consider the effects of surfaces and interfaces on phase separation processes in thin films of polymer mixtures. These surface effects can lead to striking lamellar morphologies which are easily detected using ion beam analysis techniques.

Consider a uniform blend of two homopolymers which is prepared in a one-phase region of the phase diagram and is quenched into the two-phase, unstable region of the phase diagram. For a bulk sample one expects phase separation to commence by spinodal decomposition. The mutual diffusion coefficient is formally negative, so random concentration fluctuations grow in amplitude. Long-wavelength fluctuations grow relatively slowly, because of the larger distances through which material must diffuse, while short-wavelength fluctuations are suppressed, because there is a free energy cost associated with sustaining concentration gradients which are steep on the length scale of the

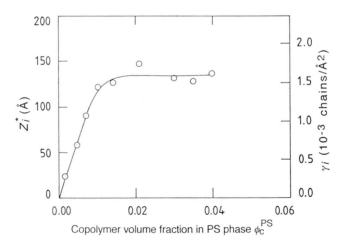

Figure 11. Interfacial excess of d-PS–PVP block copolymer at an interface between PS and PVP, as a function of bulk volume fraction of block copolymer. (Reprinted with permission from Shull *et al.*[30] Copyright (1990) American Chemical Society)

size of a polymer chain. This results in a characteristic morphology, consisting of an interconnected, random structure characterized by a single length. This may be thought of as a superposition of sinusoidal composition waves with similar wavelengths, but random phases and orientations. With increasing time, the characteristic wavelength increases as the structure coarsens. In a thin film, on the other hand, we have seen that one of the species may be preferred at the surface or interface. The effect of this is to fix the direction and the phase of the composition waves at the surface or interface producing composition waves propagating in from the surface of the film.

Such a structure is shown in Figure 12, which shows the deuterium depth profiles, as obtained using forward recoil spectrometry, for a blend of 50% poly(ethylene propylene) and 50% deuterated poly(ethylene propylene) of equal molecular weight, at various times after quenching into the two-phase region.[32] Concentrating on the vacuum surface, one sees initially segregation of the lower surface energy d-PEP at the surface and a region depleted in d-PEP behind it. As time goes on a lamellar structure develops, which coarsens with time. Note that the hydrogenous material appears to be favoured at the interface with the more polar silicon oxide. The long time structure is composed of four almost discrete layers. Detailed analysis of these data confirms that the characteristic wavelength of these structures, when extrapolated back to zero time after quenching, is in agreement with that found for bulk experiments with the same materials at the same temperatures. This correspondence is also suggested in numerical and theoretical work.[33]

A similar, though more complicated, situation arises when casting immiscible polymer blends from a common solvent. Here, as the solvent is removed, essentially one is going from a one-phase situation to a two-phase situation in the pseudo-binary system polymer A + solvent/polymer B + solvent. Figure 13 shows a depth profile obtained by helium-3 profiling for a film of a deuterated polystyrene/polybutadiene blend spin cast on to a silicon substrate from a toluene solvent.[34] This clearly shows a three-layer structure, with d-PS rich layers at the vacuum and silicon interfaces. The apparently broad interfaces between the d-PS rich phases and the PBD rich phase are in fact due to poor instrumental resolution; neutron reflectivity reveals that these interfaces have a characteristic width of only 50 Å.

4.2 Polymer–Polymer Interfaces and Diffusion

Some of the earliest applications of ion beam techniques in the polymer area were polymer diffusion measurements made by Kramer, Mayer and their coworkers. The basic experiment (Figure 14) is to take a thin film of deuterium-labelled diffusant polymer and float it on to a thicker film of protonated (normal) matrix polymer. Annealing leads to diffusional broadening of the initially sharp interface; analysis of the resulting depth profile yields a diffusion coefficient. If the matrix and diffusant are identical (apart from their isotopic labelling) this procedure yields a self-diffusion coefficient. If the matrix and diffusant are different, then, if the concentration of diffusant is low, one has a tracer diffusion coefficient. Measurement of self-diffusion coefficients and tracer diffusion coefficients for a variety of diffusants and matrices, including linear polymers of various molecular weights as well as polymers of more complicated

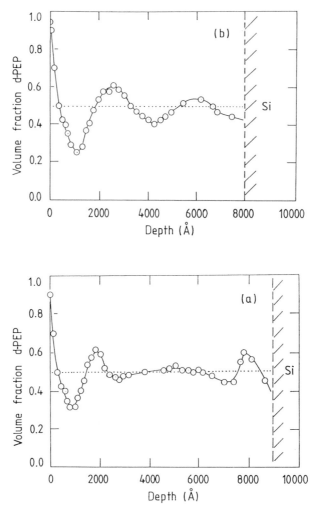

Figure 12. Forward recoil spectrometry derived depth profiles of d-PEP in a 50% blend of d-PEP and PEP, prepared as an initially uniform film and then quenched into the two-phase region of the phase diagram. (a) After 5 hours 20 minutes at 35°C, (b) after 18 hours at 35°C

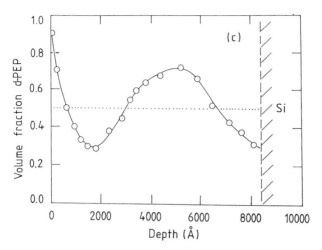

Figure 12. (c) after 2 days at 35°C. (Reproduced from Jones *et al.*[32] by permission of the American Physical Society)

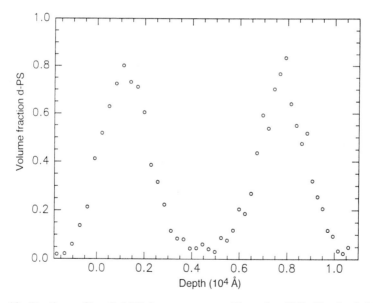

Figure 13. Depth profile of d-PS in a spun cast film of a 30% blend of d-PS in polybutadiene, as revealed by ³He profiling

Polymer Surfaces and Interfaces II

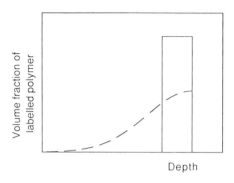

Depth

Figure 14. Measuring diffusion coefficients using ion beam scattering. A sharp interface between labelled and unlabelled polymers is created (solid line); after annealing the diffusion broadened interfacial composition profile (dashed line) is revealed by ion beam analysis

architectures such as stars and rings, has yielded many insights into the molecular mechanisms of motion in polymer melts. Ideas of reptation as well as non-reptative modes such as tube renewal have been tested. This area has been covered in a variety of reviews[10,35,36] and will not be discussed in detail here.

Additional complications are introduced if one is interested in the broadening of an interface, not between labelled and unlabelled polymers, but between chemically different polymers or polymers of widely differing molecular weight, both polymers being present in large rather than tracer concentrations (for recent reviews of this area see Kausch and Tirrell,[37] Green[10] and Klein[38]). If the two polymers are miscible in the thermodynamic sense, then the interface still broadens in a diffusive way, but the mutual diffusion coefficient that controls this broadening may well be strongly concentration dependent, leading to interfacial profiles of a non-Fickian shape. The challenge here is to relate the mutual diffusion coefficient, which controls the shape and rate of broadening of the polymer–polymer interface, to the tracer diffusion coefficients of each component, and to the thermodynamics of mixing of the polymer blend. The most complete study to date has been done on the miscible blend polystyrene–poly(xylenyl ether);[39–41] forward recoil spectrometry was used to measure tracer diffusion coefficients of each component in blends as a function of blend composition and the molecular weights of each component, to throw light on the molecular mechanisms of motion. The mutual diffusion coefficient was independently measured as a function of blend composition, and was shown to good accuracy to be given by a mole fraction weighted average of the individual tracer diffusion coefficients, multiplied by a thermodynamic factor, which enhances the mutual diffusion coefficient for systems in which mixing is thermodynamically favourable. Finally, the shape of the interface between initially pure polystyrene and pure poly(xylenyl ether) was determined using Rutherford backscattering following staining of the poly(xylenyl ether) with bromine[42], and this shape related to the known concentration dependence of the mutual diffusion coefficient. The shape of the profile was indeed highly non-Fickian; this arises from the very strong dependence of the mutual diffusion coefficient on concentration. This has two causes; the very different mobilities of the two species and the concentration dependence introduced by the thermodynamic factor. In addition, there are large additional variations in tracer diffusion coefficient as a function of blend composition that as yet are unexplained.

If the two polymers placed on top of each other have an unfavourable enthalpic interaction, yet are still miscible, then the mutual diffusion coefficient, rather than being thermodynamically enhanced, is decreased. This effect is known as 'thermodynamic slowing down'; one expects the mutual diffusion coefficient to become zero at the critical point. Such slowing down has been conclusively demonstrated in high molecular weight blends of polystyrene and deuterated polystyrene using forward recoil spectrometry to measure the mutual diffusion coefficients as a function of concentration.[43,44]

If the two polymers forming the interface are only partially miscible, on a macroscopic level no diffusion will take place for compositions inside the two-phase region. However, at a microscopic level the interface will broaden to some equilibrium width,[45,46] determined by the width at which the entropy gain of having a more diffuse interface is balanced by the enthalpy cost of the additional unfavourable contacts between the two polymers. The size of this interfacial broadening is of great importance in determining the strength of interfaces in immiscible polymer systems. Close to a critical point, the size of the interfacial region is expected to scale as

$$w \sim \frac{a}{\sqrt{\chi - \chi_c}} \tag{6}$$

where a is a polymer segment size, χ is the Flory–Huggins interaction parameter and χ_c is the value of the Flory–Huggins interaction parameter at the critical point. In most immiscible polymer blends χ is positive and much larger than χ_c; in these circumstances the interfacial width is around 50 Å, which is too small to be detected using ion beam techniques. However, in partially miscible systems, in which the critical point may be approached by adjusting the temperature, the interfacial width may become much larger. One such partially miscible polymer blend system is provided by polymethyl methacrylate– chlorinated polyethylene mixtures; in one experiment the interfacial broadening between pure chlorinated polyethylene and polymethylmethacrylate thin films was measured using Rutherford backscattering.[47] The apparent mutual diffusion coefficient greatly decreased as the critical point was approached, reflecting both the thermodynamic slowing down mentioned above and the formation of an equilibrium interface. A more well-characterized system is provided, once again, by blends of high molecular weight polystyrene and deuterated polystyrene; here the equilibrium interface width has been measured using [3]He analysis.[17,18] It was found that the variation of the width with temperature as the critical point was approached was consistent with equation (6). In addition, the dynamics of this process was followed; unusual kinetics were found.[48]

Useful information may still be gained even in the case of more strongly immiscible polymer blends, whose intrinsic interface width is less than the resolution of the technique; if a bilayer of the pure materials is made and then annealed, diffusion will take place until the equilibrium situation is reached; i.e. a bilayer of the two coexisting blend compositions will be formed. If such experiments are done as a function of temperature, the phase diagram of the polymer blend may be mapped out. Such experiments have been done in the partially miscible blend deuterated polystyrene–poly(styrene-co-bromostyrene) system, using conventional forward recoil spectrometry;[49] more recently [3]He analysis has been used in a similar way to map out the phase diagram of high molecular weight polystyrene–deuterated polystyrene blends.[50]

One of the most important technical applications of this type of work lies in relating the microscopic degree of broadening polymer–polymer interfaces to the development of mechanical strength at that interface; this is of importance for polymer welding, injection moulding and crack healing. One study directly related the development of the strength of polyimide layers to the degree of interdiffusion between them;[51] however, much work remains to be done in this area.

4.3 Penetrant Diffusion and Diffusion Limited Reaction in Thin Polymer Films

A final category concerns the transport of non-polymeric materials through thin polymer films. A prototypical case is the swelling of a glassy polymer by a small molecule solvent. It has long been known that this type of diffusion is anomalous; optical experiments reveal the presence of a sharp front separating swollen from unswollen polymer. This front moves forward with time, not proportionally to the square root of time, as in a diffusional process, but at a constant velocity. This type of swelling process is known as case II diffusion. Classical methods were unable to reveal the details of the concentration profile at the moving interface between swollen and unswollen polymers. Mills, Kramer and coworkers[7,52,53] showed that Rutherford backscattering could be used to probe the diffusion front in the swelling of various polymers, including polystyrene, polymethyl methacrylate and commercial photoresists, by halogenated solvents. In this, and later work,[54–57] many details of this process were worked out and the Thomas and Windle model was confirmed and extended. Finally,

FRES was used to extend this work to the study of deuterium-labelled hydrocarbon solvents.[58]

More complicated situations can arise when the penetrant not only diffuses into the matrix but also reacts with it. One example of this is the doping of conjugated polymers to form conducting complexes. In one study, the kinetics of penetration of AsF_5 into poly(p-phenylene vinylene) was followed using Rutherford backscattering.[59] What was revealed was a rather shallow penetration of dopant; even after 10 days exposure to vapour less than a 1 μm layer was formed. The kinetics were complicated, being neither Fickian nor case II.

A final, related situation arises from the use of curing reactions in precursor polymers. A number of polymers, which in their useful form are intractable, are prepared as high-quality thin films by spin casting a tractable precursor polymer and then curing this film to yield the desired final product. Examples are the production of polyimides such as PMDA–ODA, and the production of conjugated polymers such as polyacetylene and poly(p-phenylene vinylene). Typically the curing reaction involves the loss of a small-molecule leaving group, and it is of interest to check the progress of the curing reaction and to check that it occurs uniformly with depth. FRES has been used in one case to follow the curing reaction in PMDA–ODA polyimides.[60] The ethyl ester of PMDA–ODA, polyamic ethyl ester, was prepared with a deuterated ethanol leaving group, so that the progress of the imidization curing reaction could be followed by monitoring the loss of deuterium from the film.

5 CONCLUSION

The ion beam analysis techniques described in this chapter form a suite of mature and well-understood techniques whose potential for polymer surface and interface problems has only just started to be realized. A wide variety of problems await attack by the techniques as they currently stand. There is also considerable scope for further development and improvement of the techniques.

One area of potential development is the improvement of data collection rates; in a typical RBS experiment the detector solid angle is very small as a proportion of the total solid angle into which information yielding particles are scattered. Increasing the total area in which particles are detected could potentially greatly increase the rate of data acquisition, allowing much reduced collection times or the use of microbeams to give lateral spatial resolution. The problems, in terms of the increase in complexity of electronics and data handling, are at present large, but with the continuing development of circuit integration and the decreasing cost of computing power one can expect the situation to change over the next few years.

The use of heavy ions as projectiles[61] also presents great promise, particularly with the use of time-of-flight techniques for forward recoil spectrometry.[62–64] These techniques allow in principle the simultaneous depth analysis of a number of light elements, without interference. Again, there are software problems to be overcome; in addition one must be aware of the risks of beam damage in experiments on polymer materials.

In summary, then, ion beam techniques can now provide the polymer surface and interface community with a set of simple and well-understood analytical techniques; these are valuable either on their own or as a complement to higher resolution, but it is less easy to analyse techniques such as neutron reflectivity.

ACKNOWLEDGEMENTS

The author particularly thanks Professor Miriam Rafailovich, Professor Ed Kramer and Dr Tony Clough for stimulating and enjoyable collaborations both in the past and the present.

REFERENCES

1. Rivière, J. C., *Surface Analytical Techniques*, Clarendon Press, Oxford (1990).
2. Russell, T. P., *Mater. Sci. Rep.*, **5**, 171 (1990).
3. Kramer, E. J., Jones, R. A. L., and Norton, L. J., *Polymer Preprints*, **31**, 75 (1990).
4. Chu, W.-K., Mayer, J. W., and Nicolet, M.-A., *Backscattering Spectrometry*, Academic Press, New York (1978).
5. Doolittle, L. R., *Nucl. Instrums Meth. Phys. Res.*, **B9**, 344 (1985).
6. Doolittle, L. R., *Nucl. Instrums Meth. Phys. Res.*, **B15**, 227 (1986).
7. Mills, P. J., Palmstrom, C. J., and Kramer, E. J., *J. Mater. Sci.*, **21**, 1479 (1986).
8. Doyle, B. L., and Peercy, P. S., *App. Phys. Lett.*, **34**, 811 (1979).
9. Mills, P. J., Green, P. F., Palmstrøm, C. J., Mayer, J. W., and Kramer, E. J., *Appl. Physics Lett.*, **45**, 957 (1984).
10. Green, P. F., and Doyle, B. L., in H.-M. Tong and L. T. Nguyen (eds.), *New Characterisation Techniques for Thin Polymer Films*, John Wiley, New York (1990), p. 139.
11. Sokolov, J., Rafailovich, M. H., Jones, R. A. L., and Kramer, E. J., *Appl. Phys. Lett.*, **54**, 590 (1989).
12. Rafailovich, M. H., Sokolov, J., Zhao, X., Jones, R. A. L., and Kramer, E. J., *Hyperfine Interactions*, **62**, 45 (1990).
13. Zhao, X., Zhao, W., Sokolov, J., Rafailovich, M. H., Schwarz, S. A., Wilkens, B. J., Jones, R. A. L., and Kramer, E. J., *Macromolecules*, **24**, 5991 (1991).
14. Nagata, S., Yamaguchi, S., Fujino, Y., Hori, Y., Sugiyama, N., and Kamada, K., *Nucl. Instrums Meth.*, **B6**, 533 (1985).
15. Dieumegard, D., Dubreuil, D., and Amsel, G., *Nucl. Instrums Meth. Phys. Res.*, **166**, 431 (1979).
16. Payne, R. S., Clough, A. S., Murphy, P., and Mills, P. J., *Nucl. Instrums Meth. Phys. Res.*, **B42**, 130 (1989).
17. Chaturvedi, U. K., Steiner, U., Zak, O., Krausch, G., Schatz, G., and Klein, J., *Appl. Phys. Lett.*, **56**, 1228 (1990).

18. Chaturvedi, U. K., Steiner, U., Zak, O., Krausch, G., and Klein, J., *Phys. Rev. Lett.*, **63**, 616 (1989).
19. Jones, R. A. L., Kramer, E. J., Rafailovich, M. H., Sokolov, J., and Schwarz, S. A., *Phys. Rev. Lett.*, **62**, 280 (1989).
20. Bates, F. S., and Wignall, G. D., *Phys. Rev. Lett.*, **57**, 1429–32 (1986).
21. Schmidt, I., and Binder, K., *J. de Physique*, **46**, 1631 (1985).
22. Bartell, L. S., and Roskos, R. R., *J. Chem. Phys.*, **44**, 457 (1966).
23. Jones, R. A. L., Kramer, E. J., Rafailovich, M. H., Sokolov, J., and Schwarz, S., *Materials Research Society Symposium Proceedings*, No. 133 (1989).
24. Jones, R. A. L., Norton, L. J., Kramer, E. J., Composto, R. J., Stein, R. S., Russell, T. P., Mansour, A., Karim, A., Felcher, G. P., Rafailovich, M. H., Sokolov, J., Zhao, X., and Schwarz, S. A., *Europhys. Lett.*, **12**, 41–6 (1990).
25. Jones, R. A. L., and Kramer, E. J., *Phil. Mag. B*, **62**, 129 (1990).
26. Jones, R. A. L., Kramer, E. J., Norton, L. J., Shull, K., Felcher, G. P., Karim, A., and Fetters, L. J., *Macromolecules*, **25**, 2359 (1992).
27. Eastmond, G. C., in W. J. Feast and H. S. Munro (eds.), *Polymer Surfaces and Interfaces*, John Wiley, New York (1987).
28. Brown, H. R., *Annual Rev. Mater. Sci.*, **21**, 463 (1991).
29. Shull, K. R., and Kramer, E. J., *Macromolecules*, **23**, 4769–79 (1990).
30. Shull, K. R., Kramer, E. J., Hadziioannou, G., and Tang, W., *Macromolecules*, **23**, 4780 (1990).
31. Shull, K. R., Winey, K. I., Thomas, E. L., and Kramer, E. J., *Macromolecules*, **24**, 2748 (1991).
32. Jones, R. A. L., Norton, L. J., Kramer, E. J., Bates, F. S., and Wiltzius, P., *Phys. Rev. Lett.*, **66**, 1326 (1991).
33. Ball, R. C., and Essery, R. L. H., *J. Phys.: Condensed Matter*, **2**, 10303 (1990).
34. Geoghegan, M. A., Jones, R. A. L., Sakellariou, P., Clough, A. S., and Penfold, J., in preparation.
35. Tirrell, M., *Rubber Chem. Technol.*, **57** (1984).
36. Binder, K., and Sillescu, H., in *Encyclopedia of Polymer Science and Engineering*, John Wiley, New York (1989), p. 297.
37. Kausch, H. H., and Tirrell, M., *Annual Rev. Mater. Sci.*, **19**, 341 (1989).
38. Klein, J., *Science*, **250**, 640 (1990).
39. Composto, R. J., Mayer, J. W., Kramer, E. J., and White, D. M., *Phys. Rev. Lett.*, **57**, 1312 (1986).
40. Composto, R. J., Kramer, E. J., and White, D. M., *Nature (Lond.)*, **328**, 234 (1987).
41. Composto, R. J., Kramer, E. J., and White, D. M., *Macromolecules*, **21**, 2580 (1988).
42. Composto, R. J., and Kramer, E. J., *J. Mater. Sci.*, **26**, 2815 (1991).
43. Green, P. F., and Doyle, B. L., *Phys. Rev. Lett.*, **57**, 2407 (1986).
44. Green, P. F., and Doyle, B. L., *Macromolecules*, **20**, 2471 (1987).
45. Helfland, E., and Sapse, A. M., *J. Chem. Phys.*, **62**, 1327 (1975).
46. Helfland, E., and Tagami, Y., *J. Chem. Phys.*, **56**, 3592 (1971).
47. Rafailovich, M. H., Sokolov, J., Jones, R. A. L., Krausch, G., Klein, J., and Mills, R., *Europhys. Lett.*, **5**, 657 (1988).
48. Steiner, U., Krausch, G., Schatz, G., and Klein, J., *Phys. Rev. Lett.*, **64**, 1119 (1990).
49. Bruder, F., Brenn, R., Stühn, B., and Strobl, G. R., *Macromolecules*, **22**, 4434 (1989).
50. Budkowski, A., Steiner, U., Klein, J., and Schatz, G., *Europhys. Lett.*, **18**, 705 (1992).
51. Brown, H. R., Yang, A. C. M., Russell, T. P., Volksen, W., and Kramer, E. J., *Polymer*, **29**, 1807 (1988).
52. Mills, P. J., and Kramer, E. J., *J. Mater. Sci.*, **21**, 4151 (1986).
53. Mills, P. J., and Kramer, E. J., *J. Mater. Sci.*, **24**, 439 (1989).

54. Hui, C.-Y., Wu, K.-C., Lasky, R. C., and Kramer, E. J., *J. Appl. Phys.*, **61**, 5129 (1987).
55. Hui, C.-Y., Wu, K.-C., Lasky, R. C., and Kramer, E. J., *J. Appl. Phys.*, **61**, 5137 (1987).
56. Hui, C.-Y., Wu, K.-C., Lasky, R. C., and Kramer, E. J., *J. Electronic Packaging*, **111**, 68 (1989).
57. Lasky, R. C., Kramer, E. J., and Hui, C.-Y., *Polymer*, **29**, 673 (1988).
58. Gall, T. P., Lasky, R. C., and Kramer, E. J., *Polymer*, **31**, 1491 (1990).
59. Masse, M. A., Composto, R. J., Jones, R. A. L., and Karasz, F. E., *Macromolecules*, **23**, 3675 (1990).
60. Tead, S. F., Kramer, E. J., Russell, T. P., and Volksen, W., *Polymer*, **31**, 520 (1990).
61. Green, P. F., and Doyle, B. L., *Nucl. Instrums Meth.*, **B18**, 64 (1986).
62. Thomas, J. P., Fallavier, M., Ramdane, D., Chevarier, N., and Chevarier, A., *Nucl. Instrums Meth. Phys. Res.*, **218**, 125 (1983).
63. Thomas, J. P., Fallavier, M., and Ziani, A., *Nucl. Instrums Meth. Phys. Res.*, **B15**, 443 (1986).
64. Groleau, R., Gujrathi, S. C., and Martin, J. P., *Nucl. Instrums Meth. Phys. Res.*, **218**, 811 (1983).

5

Laser Light Scattering

J. C. Earnshaw
The Queen's University of Belfast

1 INTRODUCTION

Light scattering has long been used as a probe of structure and dynamics in supramolecular systems. Here we consider scattering by thermally excited modes of fluid interfaces[1,2] and, to a lesser extent, the associated molecular modes.[3] Fluid interfaces are continually roughened by thermal excitation. Whilst these thermally excited capillary waves are of molecular dimensions—r.m.s. amplitude is typically 2 Å for the free surface of water—they scatter light quite efficiently. This has been known since early this century, and indeed Raman investigated the scattering from the interface between a liquid and its vapour at the critical point, but until the advent of the laser it remained a scientific curiosity. The laser made possible the study of the spectrum of light scattered by the surface waves. This spectrum reflects the temporal evolution of waves of a given wavenumber, and carries information about those physical properties of the system which influence that evolution. In the presence of a molecular film at the interface, these include the viscoelastic properties of the film. As we shall see, some of the film properties potentially accessible in this way are not otherwise presently measurable.

This paper addresses the applications of surface light scattering to polymer films at liquid surfaces or interfaces. There have not been many such studies, and so it seems appropriate to devote some space to considerations of the fundamentals, paying particular attention to the potential of the technique and discussing a range of examples illustrating these themes, rather than simply reviewing the somewhat limited range of work to date. The method does hold out some unique advantages; it is not simply an alternative to other, more classical techniques.

In scattering experiments structural information usually derives from the scattered intensity, whilst the spectral content of the scattered light carries information regarding the dynamics. This broad statement remains true when the objects of study are interfacial molecular films, but further useful structural

Polymer Surfaces and Interfaces II
Edited by W. J. Feast, H. S. Munro and R. W. Richards
© 1993 John Wiley & Sons Ltd

information may be carried by the changes of the spectrum when the fluctuations of the system which scatter light are non-stationary.

The entire approach is relatively sophisticated and indirect, and it is justifiable
to ask if it is worth while. It does represent the only way to investigate certain
surface properties. Further, other techniques largely provide information on the
low-frequency response of the interfacial films, whereas the light scattering
response at high frequency may yield an insight into molecular relaxation in
the films. Even if relaxation is observed at comparatively low frequencies, the
possibility of higher frequency processes cannot be eliminated: there appears
to be a role for both classical and light scattering approaches.

2 THEORETICAL BACKGROUND

Molecular films at the interfaces between two fluids may support many modes,
from hydrodynamic motions of the system as a whole to modes of coherent
molecular fluctuation, several of which may in principle scatter light.

The dynamic interface may be analysed in terms of fluctuations which evolve
in time and space. A specific spatial perturbation can be Fourier decomposed
into a complete set of modes. Light scattering involves selection of one mode,
so we will consider the temporal evolution of a specific surface mode defined
by the interfacial wavenumber (q). The surface light scattering is dominated by
the thermally excited capillary waves. While the surface dilational waves also
cause fluctuations of dielectric constant the scattered intensity is negligible.[4]

The departure of the interface from its equilibrium plane due to a capillary
wave propagating in the x direction is given by $\zeta_0 \exp i(qx + \omega t)$. The wave
frequency ω ($= \omega_0 + i\Gamma$) represents the temporal evolution of the surface: the
spatial fluctuation oscillates with frequency ω_0 and decays at a rate Γ. The
motions associated with the capillary waves penetrate into the bulk fluid with
a decay length roughly equal to the wavelength of the surface mode, $\gtrsim 10\ \mu m$
for experimentally accessible wavenumbers ($100 < q < 2000\ \text{cm}^{-1}$). Thus, in its
influence upon the surface wave propagation, a very much thinner surface layer
can be regarded as a homogeneous film.

The theory of light scattering by molecular films is well established.[1,2,5] Two
hydrodynamic modes concern us: transverse or capillary waves on the interface,
governed by the interfacial tension (γ), and dilational or longitudinal modes,
governed by the dilational modulus (ε) of the interface. For the general case of
a liquid–liquid interface[1] the dispersion relation for the interfacial waves is
$D(\omega) = 0$ with

$$D(\omega) = [\eta(q - m) - \eta'(q - m')]^2 + \left\{ \frac{\varepsilon q^2}{\omega} + i[\eta(q + m) + \eta'(q + m')] \right\}$$

$$\times \left\{ \frac{\gamma q^2}{\omega} + \frac{g(\rho - \rho')}{\omega} - \frac{\omega(\rho + \rho')}{q} + i[\eta(q + m) + \eta'(q + m')] \right\} \quad (1)$$

where

$$m = \sqrt{q^2 + \frac{i\omega\rho}{\eta}}, \qquad \text{Re}(m) > 0 \qquad (2)$$

and η and ρ are the liquid viscosity and density (primed quantities for the upper fluid). The two physically realistic roots of the dispersion equation correspond to capillary and dilational waves.[5] However, we concentrate upon the experimentally accessible capillary waves.

In general, the two surface modes are coupled, the coupling coefficient being $[\eta(q - m) - \eta'(q - m')]$. As $\eta' \to \eta$ and $\rho' \to \rho$ the two modes become decoupled, and only then can it be correct to describe them simply as transverse or longitudinal. We therefore prefer the usage 'capillary' and 'dilational'. Due to the coupling, the temporal evolution of the capillary waves depends upon the dilational modulus as well as upon the surface tension. However, the indirectness of the influence of the dilational modulus makes its effects upon the capillary waves more subtle than those of the tension, so that ε is harder to determine than γ.

At relatively low q the capillary wave dispersion approximately follows:[6]

$$\omega_0 = \sqrt{\frac{\gamma q^3}{\rho}}, \qquad (3)$$

$$\Gamma = \frac{2\eta q^2}{\rho}. \qquad (4)$$

When both γ and ε are non-zero, coupling between the capillary and dilational waves modifies the dispersion behaviour to some extent. The major effect is a resonance between the two modes.

2.1 Surface Viscosities

The two moduli γ and ε are familiar in the low-frequency ($\omega \to 0$) limit as the interfacial or surface tension and the film compression modulus $\varepsilon = -d\gamma/d \ln \Gamma_s$, where Γ_s is the surface concentration of the molecular film. The latter modulus depends upon the stress applied: for the uniaxial case such as for capillary waves (or monolayer compression in a single barrier Langmuir trough) the dilational modulus is the sum of the bulk and shear moduli.

Dissipative effects may exist within the film. Both surface moduli can be expanded as linear response functions:[7]

$$\gamma = \gamma_0 + i\omega\gamma' \qquad (5)$$

$$\varepsilon = \varepsilon_0 + i\omega\varepsilon', \qquad (6)$$

where γ_0 and ε_0 are now the tension and elastic modulus, respectively, while the primed quantities represent surface viscosities. They are *not* the conventional surface viscosity, governing shear in the surface plane; γ' is a surface viscosity governing the response to shear transverse to the surface whilst ε' affects dilation in the surface. There is no clear consensus on the nature of these surface viscosities: they may be regarded as surface excess quantities[7,8] or as macroscopic properties of a molecular film at the interface. Neither is there any consensus on notation (see below).

These surface viscosities represent dissipative effects which influence the corresponding surface modes. Thus γ' has the effect of increasing the damping of the capillary waves compared to that due to the viscosity of the ambient liquid. Similarly ε' induces increased damping of the dilational modes.

The effects of the dilational properties upon the capillary waves are indirect, arising from the mode coupling. Figure 1 shows the effects of ε upon the capillary wave frequency and damping at a particular q. Changing ε_0 causes only slight changes in ω_0, whereas Γ is increased significantly at about $\varepsilon_0/\gamma_0 \sim 0.16$. When ε_0 is large its value has little or no effect upon the wave propagation. The dilational viscosity reduces the magnitude of the resonant interaction, until for $\varepsilon' \gtrsim 10^{-3}$ mN s/m the capillary waves become almost totally insensitive to the dilational properties. Thus if either ε_0 or ε' is large light scattering will not permit the precise determination of these properties.

The smaller coupling for an oil–water interface reduces the influence of the dilational properties even further,[9] making it particularly difficult to derive any precise values for ε_0 or ε' from light scattering experiments.

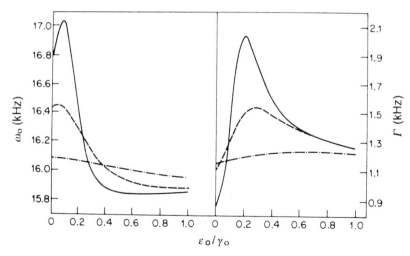

Figure 1. Variations of ω_0 and Γ with ε_0 for $q = 540$ cm^{-1}. The tension was 72.75 mN/m and the effects of three different values of ε' are shown: 0 (full line), 10^{-4} (dashed) and 5×10^{-4} mN s/m (chain)

2.2 Experimental Considerations

Quasi-elastic scattering of light can be used to investigate the behaviour of thermally excited capillary waves. The spectrum of the scattered light is just the spectrum of the thermally excited waves, which reflects their temporal evolution. It is influenced by the four surface properties discussed above, comprising two viscoelastic moduli, and can be written explicitly in terms of these moduli and the bulk properties of the subphase.[2]

Experimentally the spectrum can be measured either directly or, using photon correlation, as its Fourier transform. Neither approach seems to offer any overwhelming advantages. The latter has been used in the author's laboratories, whilst most other workers have used spectrum analysis.

Experiments are limited to relatively small q, as the intensity of scattered light falls off as q^{-2}. To measure the rather small frequency shifts of the scattered light heterodyne detection is necessary. The scattered light for a particular q is mixed at the detector with a reference beam of light at the original laser frequency, so that the detector output is modulated by the spectrum of the scattered light.

Unbiased estimates of ω_0 and Γ can be derived from such functions.[10] For a film at the air–water surface the capillary waves are affected by four surface properties, which cannot be deduced unambiguously from two measured quantities. It has been common to assume both equality of the light scattering and the equilibrium tensions and $\gamma' = 0$ to enable the data to be interpreted in terms of ε_0 and ε'. These assumptions are usually unjustified, and may be unjustifiable. In general four physical properties cannot be uniquely determined from two observables. Despite these problems, this method has been widely used in studies of monolayers.[2,11–13]

The dispersion behaviour of the capillary waves is modified by the various surface properties, so that the variation of ω_0 and Γ with q could be analysed in terms of the four surface properties. However, in at least some cases (see below) these properties are frequency dependent, vitiating this as a general approach. In fact, the four surface quantities can be directly extracted from the measured correlation functions by a fitting procedure using the analytical form of the spectrum expressed in terms of these properties.[14] Only one or two studies have used this approach (e.g. that by Earnshaw *et al.*[15]).

3 MOLECULAR FILMS

3.1 Frequency-dependent Surface Properties

The frequencies of the perturbations probed by surface light scattering are rather high—typically $10^3 < \omega_0 < 10^6 \text{ s}^{-1}$. The properties measured are not necessarily those observed at low frequencies by classical techniques. The surface properties cannot be independent of the frequency of the perturbation involved. If, say, γ' were constant then the dissipative force ($\propto \omega\gamma'$) would become infinitely large compared to the restoring force ($\propto \gamma_0$) as $\omega \to \infty$. The surface moduli must be viscoelastic. There is as yet no theory for viscoelastic relaxation in thin molecular films. However, data to date seem consistent with a simple model, the Maxwell fluid,[16] corresponding to an exponential relaxation with time constant τ.

Such viscoelastic relaxation has recently been observed for interfacial molecular films. The transverse shear modulus γ displayed relaxation for 'solvent-free' bilayers of glycerol monooleate (GMO) and for monolayers of both GMO and pentadecanoic acid (PDA). The most precise data are those from the GMO monolayer: in the fully compressed state the variations of both parts of γ were entirely compatible with a single Maxwell process with $\tau = 9\ \mu s$.[15] Plotted as $\omega_0\gamma'$ versus γ_0 as in Figure 2 the data agree well with the expected form—the semicircle shown. Note that γ_0 tends to the measured $\gamma_0(eq)$ as $\omega \to 0$, showing that there was no relaxation slower than that cited.

The other two systems mentioned behaved similarly. For PDA monolayers[17] below the triple point the relaxation time was about 20 μs. While somewhat scattered the bilayer data[18] were consistent with a Maxwell relaxation with $\tau = 37\ \mu s$. The T dependence of the relaxation time for the latter case could be estimated, τ being lowest just at the membrane transition.[19]

These findings are all consistent with acyl chain melting. The time scales are compatible with values from temperature jump studies of bilayer transitions.[20]

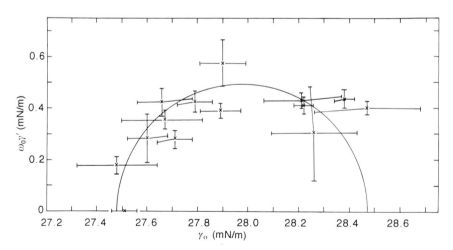

Figure 2. A Cole–Cole plot of the transverse shear modulus γ for a fully compressed GMO monolayer. The semicircle represents the behaviour of a Maxwell relaxation of strength 1 mN/m and time scale 9 μs

In the bilayer the apposed lipid films would sterically hinder straightening of the acyl chains, causing the larger τ value. Similarly, chain melting for PDA would be inhibited by the more closely packed fully saturated chains than the unsaturated oleate chains of GMO. Such chain melting leads to bobbing up and down of the lipid molecules, which has been associated with passive ionic permeation through membranes.[21] The quicker relaxation at the bilayer transition is entirely compatible with the established transition peak in lipid vesicle permeability.[22]

The association of the film viscosity γ' with chain–chain interactions is supported by data from lipid bilayers. For those formed from solutions of GMO in decane, which are known to retain a large quantity of solvent in their structure, γ' was negligible, whereas 'solvent-free' bilayers displayed finite γ'.[23] The incorporated decane would enable the acyl chains to move freely, whereas in 'solvent-free' bilayers the steric resistance to chain flexing would be increased.

Relaxation of the dilational modulus ε occurs within some regions of the phase diagram in all monolayers that have been studied.[15,17] For example, for PDA above the triple point, the elastic modulus ε_0 was non-zero in the coexistence regions, where the flatness of the classical isotherms implies that $\varepsilon_0(\text{eq}) = 0$. The relaxation processes involved appear to be comparatively slow ($\tau \ll 1/\omega_0$), so that the viscosity ε' was commonly very low at light scattering frequencies. This, together with the indirect action of ε upon the capillary waves, made it difficult to observe frequency dependence of ε.[24]

Recalling that the modulus ε as derived from light scattering is the sum of those for shear and pure compression, its relaxation might involve several processes. In only one case have comparisons been possible: just at the condensed end of the LC/LE transition of PDA the light scattering ε' seemed to comprise the shear component alone, the dilational viscosity having relaxed entirely into the storage modulus ε_0.[17] Any relaxation involving the shear viscosity is much faster than $1/\omega_0$.

3.2 Non-stationarity of the Light Scattering Response

The surface properties recovered by light scattering are *local* averages over the area illuminated by the laser beam, of dimension D. Inhomogeneities in an interfacial film affect the observed correlation functions. If the typical domain size d is larger than D the monolayer will appear instantaneously homogeneous, and the light scattering signal may fluctuate between those appropriate to the different phases. However, if $d < D$ the light scattering signal will be a weighted average of those from the separate phases, reducing any fluctuations.

Fluctuations have been observed for monolayers and ascribed to coexistence of phase separated domains.[11–13] However, the usual long experimental durations would have temporally averaged any fluctuations. Minor changes in the

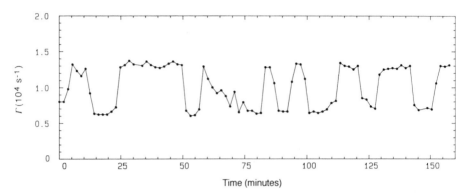

Figure 3. Capillary wave damping as a function of time for a PDA monolayer at 19°C at a surface concentration of 0.017 molecules/Å. The fluctuations indicate the coexistence of liquid and vapour domains in the monolayer

operation of our apparatus enabled very rapid (~ 10 s) acquisition of statistically adequate correlation functions,[25] enabling the temporal averaging to be reduced by at least an order of magnitude.

The simplest system to consider is the monolayer at the air–water interface. PDA is a particularly suitable material for such studies as the transitions of the molecular film are simple first order in nature.[26] We show only one example.

In the liquid–vapour transition of PDA successive correlation functions fluctuated visibly.[27] In the vapour phase the surface resembled the clean subphase, whereas in the liquid phase ε_0 (~ 13 mN/m) was close to the value for resonance (Figure 1). Thus Γ was the natural observable. Its fluctuations, shown in Figure 3, range between the two extreme values, establishing that the entire illuminated area could be occupied by one of the phases at a time. The data suggest that the liquid domains are about 1.1 ± 0.7 cm in size. For the liquid expanded–liquid condensed transition the fluctuations were smaller: the LC domains were $\sim 500 \ \mu$m across. These domain sizes, although an order of magnitude greater than suggested by fluorescence microscopy methods,[28] agree well with values deduced from surface potential fluctuations.[29]

3.3 Information from the Intensities

The *relative* intensities of the scattered light (I_s) and of the heterodyne reference beam (I_r) can be extracted from the correlation functions.[30,31] While absolute values of the intensities cannot be obtained, variations in their magnitudes can be followed and have proved informative in some cases. This method of determining the scattered intensity is rather direct, as it automatically excludes extraneous scattering processes of different time dependence. The intensity scattered by thermally excited capillary waves is[4]

$$I_s \propto R \frac{kT}{\gamma_0 q^2} \tag{7}$$

where R is the interfacial reflectivity. Thus $I_s(q)$ reflects only the thermal origin of the capillary waves, carrying no structural information. The q and γ_0 dependence of I_s for monolayers agreed well with these predictions;[31] the scattered intensity is indeed just that due to the capillary waves.

We have used the temperature variations of both I_s and I_r to infer changes in film structure. The most illuminating example comes from a study of the thermotropic transitions of 'solvent-free' bilayer membranes of GMO.[32]

The I_r data, shown in Figure 4, changed considerably about the transition temperature, and displayed rather pronounced fluctuations below T_t. The peak between 12 and 14°C is apparently unconnected with structural changes, and will be discounted here. The reference beam was reflected at the bilayer, so $I_r \propto R$, which depends upon the membrane thickness as h^2. Thus I_r can be interpreted directly in terms of the variation of h^2, given that the mean refractive index of the bilayer is independent of T. Using a literature datum[33] for normalization, the I_R scale can be converted to one of h (see Figure 4). The

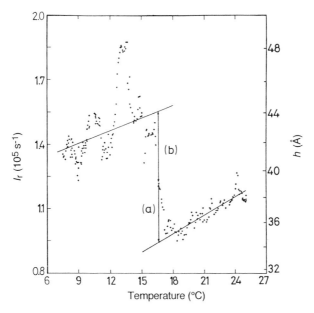

Figure 4. The T variation of I_r. Linear fits above and below T_t are shown (neglecting the peak between 12 and 14°C). The non-linear right-hand scale indicates r.m.s. bilayer thickness

bilayer is 44 Å thick just below the transition, in good accord with expectation for a bimolecular film of fully extended GMO molecules oriented normal to the film. The decrease of 8 Å through the transition is compatible with some 6 or 7 *gauche* rotations per GMO molecule.

This change in membrane thickness was exactly paralleled by a small change in tension over a rather narrow range of T.[32] This rather convincingly supports an identification of the transition as involving chain melting: below the transition the lipid acyl chains tend to adopt an all-*trans* configuration normal to the bilayer; the consequent increase in packing density at the water–bilayer interfaces led to a reduction in tension.

The scattered intensity for this bilayer is shown in Figure 5. Equation (7) shows that the scattering by thermally excited capillary waves depends upon R and γ_0. The T dependence of both these quantities was known—R via I_r and γ_0 from the wave frequencies. Combining the *forms* of the two variations as in equation (7) led to predictions for I_s shown by the lines in Figure 5. There are some uncertainties in the functional dependence of γ_0 in the region of the transition itself, so the line is dashed there.

Above 18°C the agreement of the experimental I_s data with the predicted form is excellent. The substantial discrepancies below this temperature require explanation. Any processes invoked to explain the discrepancies of Figure 5

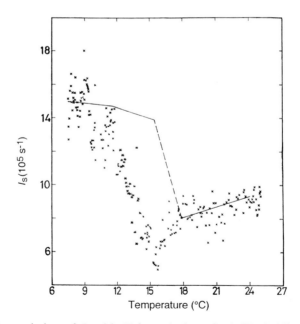

Figure 5. The variation of I_s with T for a 'solvent-free' GMO bilayer. The lines, normalized to the data at 24°C, are discussed in the text

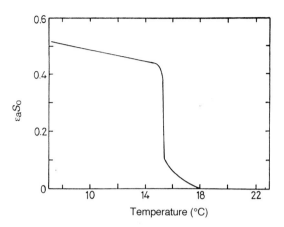

Figure 6. The variation of dielectric anisotropy multiplied by the lipid orientational order parameter inferred from the data of Figure 5

must be closely coupled to the capillary waves, because the time-dependent part of the correlation functions, from whose amplitude I_s derives, agreed excellently with the form expected for capillary waves on the bilayer. Some intramembrane molecular modes couple to the hydrodynamic modes of interest and may scatter light.[3] In particular, molecular splay couples to the capillary waves. The effects on the capillary wave propagation, arising from $\gamma \to \gamma + Kq^2$, are negligible as the splay modulus $K (\sim 10^{-19}$ J) is far too low to affect experimentally accessible wavenumbers (< 2000 cm^{-1}). However, the light scattered by the molecular modes may interfere with that scattered by the capillary modes.[3] The intensity scattered by the splay modes depends upon the combined dielectric anisotropy and chain order parameter of the bilayer ($\varepsilon_a S_0$).[3] The observed I_s behaviour implies a rather abrupt increase in this combination (Figure 6) at the transition. The scattered intensities thus supported the conclusions from I_r: there is a substantial and abrupt change in the order of the lipid molecules just when the thickness becomes equal to two fully extended GMO molecules. The entire picture is satisfyingly coherent.

4 POLYMER FILMS

There have been comparatively few reports of experimental investigations of surface or interfacial polymer films by laser light scattering. Apart from one or two cases where polymers were studied as part of a more general investigation of monolayer viscoelasticity,[2,13] these reports have emanated from the laboratory of H. Yu and coworkers. We focus our discussion on this more coherent body of work.

A considerable range of studies has been undertaken.[34-40] Different materials have been studied, covering a range of hydrophobicity;[35] the difference between monolayers of a polymer and its monomer has been investigated:[36] spread and adsorbed monolayers have been compared,[37] as have homopolymers and block copolymers,[39] and the temperature dependence of the surface viscoelasticity of a particular polymeric material has been studied.[40] Space precludes a survey of the entire corpus of work, so we highlight a few particular features.

Polymer films at fluid interfaces may differ from the bulk in that the chains may be mobile and may act almost as a viscous liquid. The surface mobility may well vary considerably from polymer to polymer, presumably being related in some way to the glass transition of the bulk material. Segregation of the components of block copolymers into the different fluid phases may modify the mechanical properties.

The light scattering experiments used spectrum analysis in the frequency domain. Spectra required about 1 minute to record, so some degree of temporal averaging must have occurred. The intensities were not determined.

Typical spectra recorded for the surface of water and for a spread monolayer of poly(vinyl acetate) (PVAc) are shown in Figure 7.[34] The monolayer clearly causes a decrease in the central frequency of the spectrum for each wavenumber, as well as an increase in the spectral width. The spectra were not analysed directly in terms of the surface properties but were fitted by a Lorentzian form to determine the frequency (f_s) and damping $(\Delta f_{s,c})$ of the capillary waves. The fitted linewidths were corrected for instrumental line-broadening effects.

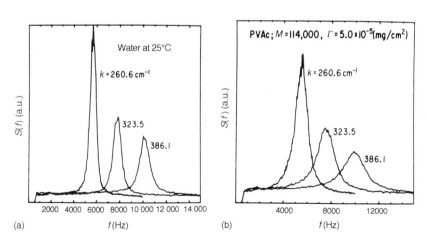

Figure 7. Power spectra for three different wavenumbers recorded for clean water and for a PVAc (molecular weight 114 000) film of surface concentration 5×10^{-8} g/cm^2. (Reprinted with permission from Kawaguchi *et al.*[34] Copyright (1986) American Chemical Society)

Table 1. Comparison of the notations used by different workers in surface light scattering.

Quantity	Belfast	Wisconsin
Wavenumber	q	k
Frequency	ω_0	$f_s = \omega_0/2\pi$
Damping	Γ	$\Delta f_{s,c} = \Gamma/4\pi$
Tension	γ_0	σ
Transverse shear viscosity	γ'	μ
Elastic modulus	ε_0	ε
Dilational viscosity	ε'	κ
Viscoelastic moduli	γ	σ^*
	ε	ε^*

There is no agreed notation in this field, and to avoid confusion we switch in the remainder of this review to the notation used by the original authors. To assist the reader the different symbols are listed in Table 1. Henceforward Γ will indicate surface concentration of film molecules.

The basic data from the light scattering experiments were the f_s and $\Delta f_{s,c}$ values at each k observed. These were used to estimate the surface viscoelastic parameters by solving the dispersion equation for ε and κ. This requires extra information concerning the other surface properties: it was usually assumed that the tension equalled the value measured with a Wilhelmy plate and that the transverse shear viscosity $\mu = 0$. The fact that these assumptions led to consistent values of ε and κ was their only justification (see above). The present discussion will rest upon the values of ε^* thus found, bearing in mind the caveats regarding their derivation.

4.1 Studies of Different Materials

The equilibrium and dynamic surface properties of films of six polymers of varying hydrophobicity were studied.[35] These included poly(ethylene oxide) (PEO), polytetrahydrofuran (PTHF), PVAc, poly(methyl acrylate) (PMA), poly-(methyl methacrylate) (PMMA) and poly(*tert*-butyl methacrylate) (PtBMA). The first four formed liquid monolayers, whereas the last two form condensed monolayers. The latter are less amenable to surface light scattering, which is less sensitive to high values of ε. The static properties of the monolayers were relatively similar; it was hoped that differences between the high-frequency viscoelastic behaviour might provide some insight into molecular dynamics.

The capillary wave frequencies and damping values are shown in Figure 8 for all of these materials. In all cases f_s remains constant over a range of low surface concentrations, whereas in some cases the initial increase in $\Delta f_{s,c}$ occurred at rather low Γ. The frequency basically tracks the changing tension,

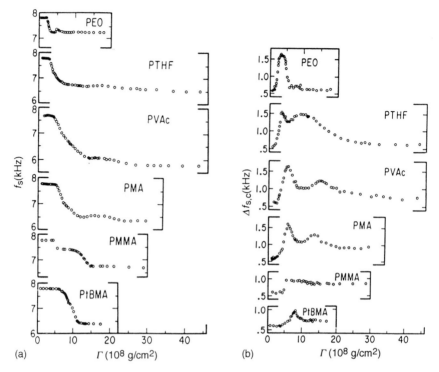

Figure 8. Frequencies and damping values as a function of surface concentration for six polymers at $k = 323$ cm^{-1}. Data for other wavenumbers are not shown. (Reprinted with permission from Kawaguchi *et al.*[35] Copyright (1989) American Chemical Society)

whilst the variations of $\Delta f_{s,c}$ reflect the interplay between σ and ε, modulated by the dilational viscosity (assuming that $\mu = 0$). One of the condensed monolayer materials (PMMA) exhibited a rather different variation of both f_s and $\Delta f_{s,c}$. These frequencies are not the main results, and so we dwell no further upon them.

The viscoelastic properties of the monolayers are summarized in Figures 9 to 11. The scatter of data from several surface wavenumbers generally indicates the reproducibility of these properties.

4.1.1   PEO and PTHF

The viscoelastic behaviour of PTHF closely resembled that of PEO (Figure 9). In both cases, the static and dynamic estimates of ε agreed up to about 10 mN/m, but for more concentrated films the light scattering values exceed the classical data. The discrepancy was larger for PTHF (for which ε reached

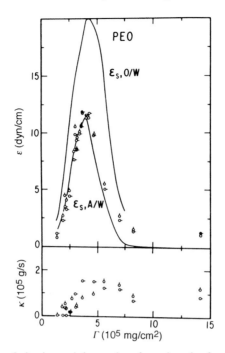

Figure 9. The dilational elastic modulus and surface viscosity for PEO. The lower solid line derives from the equilibrium tension data, while the points indicate the light scattering values, derived as described in the text. Data from two wavenumbers are shown. The solid line labelled O/W refers to the heptane–water interface and is discussed in the text. (Reprinted with permission from Sauer *et al.*[39] Copyright (1987) American Chemical Society)

values about twice those for PEO) than for PEO, being about 5 mN/m at the largest in the former case, compared to 2 mN/m for PEO. The data thus suggest that viscoelastic relaxation occurred in the more concentrated monolayers, with relaxation time $> 1/2\pi f_s$. The discrepancies between the static and dynamic values of ε appeared at lower area/monomer than the limiting value. In this regime the monolayer will increasingly depart from a quasi two-dimensional system by chain looping or macroscopic film collapse, and relaxation of part or all of the chains seems entirely plausible.

The dilational viscosity κ also behaved similarly in both cases, rising from small values, reaching a maximum at the same point as did ε and falling to a plateau at larger Γ. The coincidence of the maxima in κ and ε seems general, occurring for all polymers studied. It has no obvious explanation. Again, the values of κ for PTHF were greater than those shown for PEO. The larger values of both ε and κ for PTHF presumably reflect its greater hydrophobicity, resulting in increased intra-molecular interactions.

Close inspection of the data shows that for both materials κ appears to fall with increasing k. The effect seems to be limited to that range of Γ at which the light scattering ε exceeds its static counterpart. This suggests the reality of the effect, as the viscous part of a viscoelastic modulus should fall with increasing f_s above the inverse of the relaxation time, where the elastic part is increased.

4.1.2   PVAc and PMA

PVAc and PMA form rather more condensed monolayers than the previous pair. Structural similarities between the two suggest that their behaviour should not differ very greatly, as is the case (Figure 10). The elastic moduli were somewhat greater than for the polyethers, and again the light scattering data only agreed with equilibrium values to well below the limiting area per monomer. Thereafter the dynamic values systematically exceeded the static values by a few millinewtons per metre. The dilational surface viscosity peaked where ε maximized, but also rose significantly at larger Γ, probably reflecting greater chain–chain interactions as the segments were forced out of the surface. The difference from the polyethers presumably arises from the greater hydrophobicity of the present materials. Further, PVAc and PMA are close to their glass transitions at the experimental temperature of 25°C, where the polyethers are well above their transitions.

4.1.3   PMMA and PtBMA

Both PMMA and PtBMA form highly viscoelastic monolayers: both ε and κ were significantly larger than for the other materials (Figure 11). Neither ε nor κ can be accurately determined when their magnitudes are so large (cf. Figure 1). PMMA differed significantly from all the other materials, in that both ε and κ increased at Γ well below the lift-off of the classical $\pi - A$ isotherm. This material is the only one that exhibits biphasic behaviour: in the dilute monolayer small but visible patches of condensed material are visible. Fluctuations of the light scattering signal were not reported, so the domains were not large enough to cover the illuminated spot. Such condensation in dilute films implies extremely strong intermolecular cohesion. The very large surface viscosities would follow, as dilational stress would lead to considerable dissipation of energy. Alternatively, the array of domains, whilst mobile under slow stress, might resist dilational stress at high frequencies, presenting the appearance of an incompressible network.

PtBMA again differed from the other materials studied in that ε became very large at Γ above that corresponding to the peak in the classical ε. However, when κ is so large (Figure 11) it is difficult to determine ε with any very great precision.

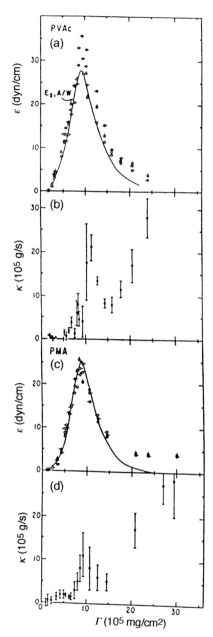

Figure 10. As for Figure 9, but for PVAc and PMA. The light scattering data are for three wavenumbers. (Reprinted with permission from Kawaguchi *et al.*[35] Copyright (1989) American Chemical Society)

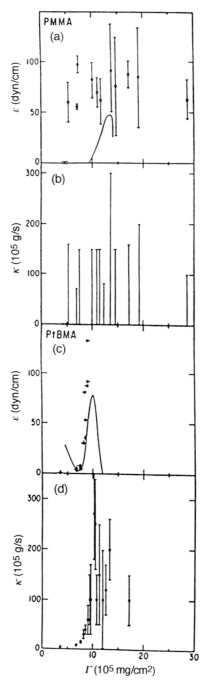

Figure 11. As for Figure 9, but for PMMA and PtBMA. For the latter, ε becomes very large when Γ exceeds 10×10^{-5} mg/cm^2. (Reprinted with permission from Kawaguchi *et al.*[35] Copyright (1989) American Chemical Society)

In summary, this investigation demonstrated the ability of light scattering to differentiate polymers by their viscoelastic behaviour. In most cases it appeared that viscoelastic relaxation must have taken place, particularly in denser films. The overall trend of increasing dilational surface viscosity with increasing polymer hydrophobicity may arise from increasing segment cohesion.

4.2 Films of Vinyl Stearate and Poly(Vinyl Stearate)

An interesting study[36] contrasted the surface light scattering from spread monolayers of poly(vinyl stearate) (PVS) and of its precursor monomeric material, vinyl stearate (VS) at the air–water interface. The goal was to compare the viscoelastic properties of two materials having the same basic chemical unit, to gain some insight into the influence of restrictions of chemical packing upon these properties.

The basic light scattering data are shown in Figure 12; it immediately confirms that the two materials differ significantly. It has to be said that there is a minor problem with the PVS data: at large areas per monomer the $\Delta f_{s,c}$ data consistently fell below those for the clean subphase. This cannot arise from monolayer viscoelasticity, which tends to *increase* the damping of the capillary waves, but apparently represents some technical problem. It may be that the correction for the instrumental effect was slightly too large in this case. At all events the effect does not modify the principal conclusions to be drawn from the data.

On monolayer compression the behaviours of the capillary waves for the two materials were completely different, most evident for $\Delta f_{s,c}$. Whereas changes in f_s and $\Delta f_{s,c}$ for VS only occurred as the monolayer was compressed below 30 Å^2/molecule, such changes were apparent in PVS monolayers as expanded as 55 Å^2 monomers. In both cases the damping changed at rather larger molecular areas than f_s. The wave frequencies changed for VS just where the monolayer surface pressure began to increase, but for PVS the changes occurred at much larger areas than the lift-off of the $\pi - A$ isotherm. The drop in f_s for PVS seemed to occur just where $\Delta f_{s,c}$ reaches its maximum. This maximum differed significantly in form from those observed for the materials considered above, being smaller and broader.

These basic data were interpreted in terms of the surface viscoelasticity, under the assumptions discussed above. The variations of ε and κ thus found are shown in Figure 13, together with the classically determined $\pi - A$ isotherms.

The surface dilational modulus remained essentially zero for VS until the surface pressure departed from zero. However, for PVS both ε and κ increase sharply at about 55 Å^2/monomer. The apparently finite surface viscosity for both materials at large areas is probably compatible with a value of zero, but the increases at lower areas appear real; $\kappa \sim 10^{-3}$ mN s/m is sufficiently large to suppress the resonance between capillary and dilational waves, rendering light

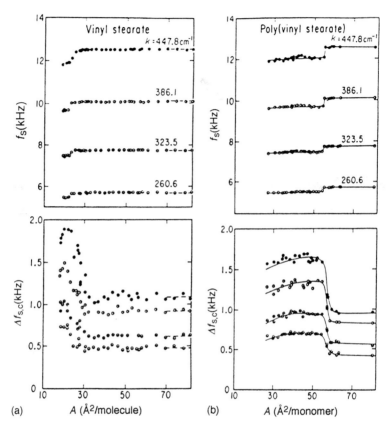

Figure 12. Capillary wave frequencies and damping values for VS and PVS for four different wavenumbers. The dashed horizontal lines on the plots for the monomer represent the values expected for the free subphase surface. (Reprinted with permission from Chen *et al.*[36] Copyright (1987) American Chemical Society)

scattering somewhat insensitive to ε (cf. Figure 1). We thus conclude that the apparent decrease of ε for PVS on compression to areas less than 50 Å^2/monomer may not be a real effect. However, the dilational properties of the polymer clearly differ significantly from those of the monomer. It should be emphasized that these effects were only revealed by the light scattering observations; the static $\pi - A$ isotherms (and thus the equilibrium ε) gave no hint of them.

It would be premature to speculate on molecular mechanisms for these differences in dilational properties for PVS and VS. Only two of the materials discussed above (PMMA and PtBMA) showed surface viscosities comparable with those for PVS, so that the polymeric nature of PVS is not the only reason for the differences from VS. It cannot just be a question of a slow viscoelastic

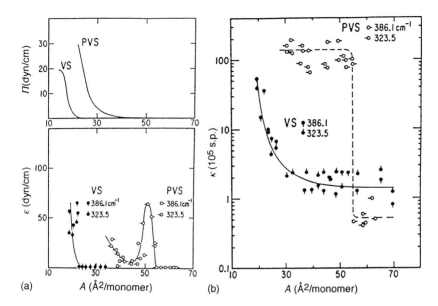

Figure 13. (a) $\pi - A$ plots and (b) the surface properties for VS (filled circles) and PVS (open circles) as functions of area per molecule or per monomer unit. (Reprinted with permission from Chen *et al.*[36] Copyright (1987) American Chemical Society)

relaxation due to steric effects and/or chain dynamics: at capillary wave frequencies above the inverse of the relaxation time the surface viscosity κ would be expected to fall rapidly with f_s, whereas there is no indication that this is the case (Figure 13). In fact the polymer film is considerably more viscous than the monomeric one: the loss modulus, $\omega\kappa$, is much larger relative to the storage modulus ε for PVS than for VS at all molecular areas. It may be that there are two relaxation processes involved, one (of time scale $> 1/2\pi f_s$) responsible for the difference between the light scattering and equilibrium values of ε and one (much faster) causing κ to be non-negligible at frequencies well above the first relaxation.

4.3 Spread and Adsorbed Films

PEO is water soluble, but its high surface activity permits the formation of stable surface films. Stable films can also be spread at the air–water interface. PEO apparently adopts a flattened, quasi two-dimensional structure at the surface with most segments in contact with the surface. The process of adsorption can be seen as cooperative, the stability of the film arising from the large number of contacts with the surface, each segment being only slightly amphiphilic. Spread films might differ from films formed by adsorption from solution, as the segments attached to the surface might be in different physical states. Light scattering offers the possibility of investigating such differences, at least insofar as they affect the film dynamics. Experimentally[37] the films formed by the two methods appeared indistinguishable: the data (f_s and $\Delta f_{s,c}$), measured as a function of surface pressure, were found to be identical within errors for both cases.

The comparison between the viscoelastic properties of spread and adsorbed films would have been difficult without light scattering. Equilibrium methods for determination of ε for the adsorbed case would require knowledge of the $\pi - \Gamma$ isotherm, whereas the light scattering is directly sensitive to this quantity (albeit its high frequency value), as well as to the surface viscosities κ and μ. It seems highly improbable that f_s and $\Delta f_{s,c}$ could agree accidentally for two arbitrary sets of the four surface viscoelastic properties. The conclusion that the surface properties of the films are identical seems safe. PEO forms molecularly similar films by both routes over the entire range of π observed (≤ 10 mN/m). The activation barrier for adsorption is apparently rather small. The independence of film state upon the path of formation was used to determine surface concentration as a function of time in the adsorbed films from the measured π, in a study of the adsorption dynamics.[37]

In general, for soluble surfactants, diffusive exchange between surface film and bulk solution leads to a specific constitutive equation describing the frequency dependence of ε^*, scaled by a characteristic time scale for the exchange.[5] While this theory was developed for small molecule surfactants, it presumably would apply, *mutatis mutandis*, to the polymer case. The present data, for which ε^* for the adsorbed film was identical with that for spread film, suggests that the time scale for diffusive exchange of PEO must be much longer than the inverse of the capillary wave frequency ($\omega_0 \geq 4.4 \times 10^4 \text{ s}^{-1}$ [37]).

4.4   Films at the Air–Water and Oil–Water Interfaces

Comparatively minor modifications of the light scattering apparatus permitted observation of scattering from a liquid–liquid interface, specifically heptane–water.[38,39] Comparisons of identical spread polymer films at air–water (A/W) and oil–water (O/W) interfaces help to point up the advantages and disadvantages of the technique for the latter case.[39]

The differences result from the reduced coupling between the capillary and dilational surface waves when the densities and viscosities of the two fluids approach each other. The coupling is much less significant for O/W than for A/W. As the effects of the coupling are solely responsible for the sensitivity of light scattering to the surface dilational properties, we may expect this sensitivity to be drastically reduced for a fluid–fluid interface. Whereas the variation of $\Delta f_{s,c}$ from its value for clean water may reach 200% for A/W (Figure 1), it only reaches 8% for heptane–water. Similarly, f_s, instead of varying by some 6%, changes by less than 1%.

While it is not strictly correct to neglect the effects of the coupling, they are very small. In particular, the potential variation of f_s is so small that the assumption that this frequency is purely due to interfacial tension cannot be much in error. Light scattering thus provides a non-invasive technique for the measurement of interfacial tension. However, the sensitivity to the surface dilational properties is so low that light scattering hardly provides a viable method of investigation.

These points may be illustrated by light scattering data for PEO[39] at the heptane–water interface (Figure 14) which display completely different variations from those for A/W (Figure 8). The O/W data were interpreted in terms of the surface pressure and the transverse shear viscosity (μ), neglecting the coupling between the surface modes. The dispersion equation was solved, using the experimental f_s and $\Delta f_{s,c}$ and assigning values (unspecified—perhaps the equilibrium values?) for ε and κ, to yield σ^*. The π values thus determined agreed with Wilhelmy plate data, strengthening the arguments that the wave

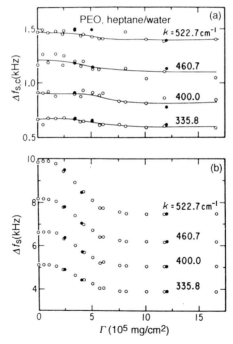

Figure 14. Capillary wave frequencies and damping constants for PEO at the oil–water interface for various k values. Air–water data can be found in Figure 8. (Reprinted with permission from Sauer *et al.*[39] Copyright (1987) American Chemical Society)

frequencies are almost entirely determined by σ for O/W. It was concluded that μ was compatible with zero: the observed variations of $\Delta f_{s,c}$ are apparently due to ε^*, via the residual coupling between the surface modes.[38] The transverse shear viscosity may indeed be zero. The oil phase may act to fluidize the hydrophobic chain segments, as is apparently the case for lipid bilayers.[23] Bilayers containing significant amounts of decane in their structure appear to have no transverse shear viscosity, whereas 'solvent-free' BLM exhibit finite values for μ.

The static values of ε for PEO at the O/W interface (Figure 9) varied with Γ in the same manner as for A/W, but were roughly twice as large in magnitude, whereas for PVAc the two quantities were similar in size. It appears that the chain interactions of PEO at the O/W interface are stronger than for A/W.[39] The difference from PVAc, for which the two cases were similar, presumably reflects the more hydrophilic nature of the PEO segments.

4.5 Block copolymers

A low molecular weight poly(ethylene oxide)–polystyrene block copolymer, constituting a non-ionic surfactant, was investigated at both air–water and heptane–water interfaces.[38] The polymer could be spread at both interfaces, PS

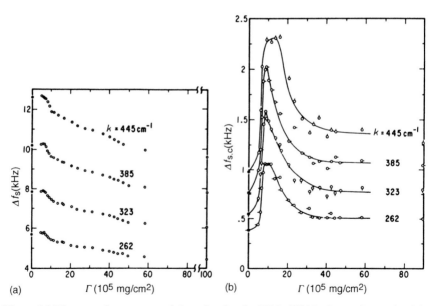

Figure 15. The wave frequency and damping for the PEO–PS block copolymer for A/W. The filled circles indicate the values of the clean water surface. (Reprinted with permission from Sauer *et al.*[38] Copyright (1987) American Chemical Society)

endowing segments of the chain with considerable hydrophobicity. Light scattering was only undertaken over a restricted range of Γ, above which π exhibited hysteresis, perhaps because of multilayer formation or desorption.

The light scattering data for the A/W case are shown in Figure 15. Whereas f_s remained roughly constant over a finite range of Γ, $\Delta f_{s,c}$ rose towards the resonant maximum from the lowest concentrations used. These data were analysed in terms of the dilational modulus ε^*, assuming that π equalled or exceeded its static value and that $\kappa \geq 0$ and $\mu = 0$. The values of ε and κ thus determined varied with π, both being maximal when π equalled its equilibrium value. Permitting the surface *pressure* to exceed the static value seems slightly strange, as any viscoelastic relaxation would affect σ directly, rather than π. Indeed, in later work (see above) π was always held constant *at* the static value. We thus concentrate upon the upper bounds of the dilational properties of Figure 16.

As for most homopolymers investigated, the static and light scattering values of ε were similar in magnitude, and follow broadly similar courses with Γ. The results did not differ greatly from those observed for PEO (Figure 9), although

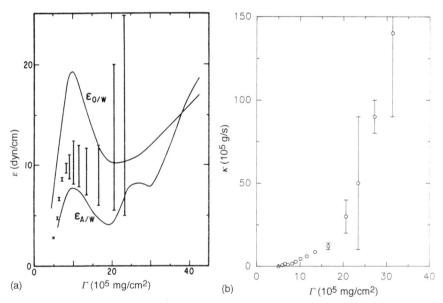

Figure 16. Dilational properties of the PEO–PS block copolymer. (a) The full lines show the variations of the static values of ε for the A/W and O/W cases. The vertical lines indicate the range of values determined from the light scattering data as described in the text. (b) The corresponding κ data for the A/W case. The points correspond to the tops of the vertical lines in (a). (Reprinted with permission from Sauer *et al.*[38] Copyright (1987) American Chemical Society)

there seems rather more difference between the two sets of f_s and $\Delta f_{s,c}$, the difference appearing to grow with the magnitude of κ. The surface viscosity was considerably larger than that observed for PEO, for which $\kappa \leq 3 \times 10^{-5}$ mN s/m. In particular, the copolymer became much more viscous, relative to the elastic modulus ε, at larger Γ. This presumably arose from increased chain–chain interactions, ascribed to PS segments as the PEO segments were forced into the water phase.

For the heptane–water case the light scattering data behaved very differently from the A/W case (Figure 17). The frequency variation simply reflected the change of σ with Γ, while $\Delta f_{s,c}$ varied only slightly, due to the reduced coupling. As for PEO at the O/W interface, the data were consistent with $\mu = 0$. The value of ε found from the $\pi - A$ behaviour is indicated in Figure 16; no attempt was made to determine either ε or κ from the light scattering data. The ε data agreed quite well, in magnitude and in trend, with those found for PEO at the O/W interface.

Thus, despite apparently different variations of $\Delta f_{s,c}$ for the copolymer compared to PEO, the elastic modulus was found to behave similarly in both cases. (For the A/W case the large κ for the copolymer would have affected f_s and $\Delta f_{s,c}$ quite considerably.) The point is that the different behaviours of the copolymer at the A/W and O/W interfaces appear to arise solely from the PEO

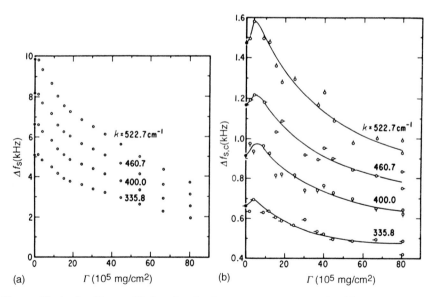

Figure 17. As for Figure 15, but for the heptane–water case. The filled circles again represent the properties of the clean interface. (Reprinted with permission from Sauer *et al.*[38] Copyright (1987) American Chemical Society)

segments. It was speculated that this is because in both cases PEO dominates the interface itself, determining the area per molecule for $\Gamma < 15 \times 10^{-5}$ mg/cm^2. In contrast it appears that PS must play a greater role at surface concentrations $> 20 \times 10^{-5}$ mg/cm^2, where π rose well above the collapse point of PEO. This corresponded to the onset of the rapid rises of both κ and ε. These changes may be due to the displacement of PEO segments from the interface and into the water phase. At these areas each copolymer molecule occupies a substantial area, indicating that PS segments are still at the interface (either A/W or O/W).

5 CONCLUSIONS

In conventional light scattering the dynamic structure factor $S(q, \omega)$ yields information on both structure and dynamics: structure through its q dependence and dynamics through the ω variation. In the present case the q^{-2} variation of the scattered intensity carries no structural information, apart from the decreasing amplitude of the thermally excited waves of smaller wavelength. The motions of the surface or interface are, of course, reflected in the frequency dependence of the scattered spectrum. However, as we have seen, much more information is potentially available from the light scattering data.

Much of this extra information only becomes available when the experiment, and the analysis of the data, are pushed to the limits. Thus structural information can be carried by fluctuations in the detected spectrum; without rapid data acquisition techniques[25] such fluctuations will be reduced by averaging. Similarly the scattered and heterodyne reference beam intensities can, in some circumstances, carry useful information,[32] which is not available if the data are not analysed in this way. Finally, details of intrafilm molecular relaxation are carried by the frequency dependence of the surface viscoelastic properties.[15] The observation of such relaxation requires, firstly, that a fairly wide range of wavenumber should be studied and, secondly, that the surface properties should be extracted from the light scattering data as precisely and unambiguously as possible. The latter requires, in practice, an approach to data analysis along the lines of the direct method mentioned above.[14]

Quasi-elastic light scattering has proven a useful tool in the investigation of molecular films. Much detail has been determined in studies of the thermotropic transitions of both lipid bilayers and insoluble monolayers. This has included observation of phase separation and of changes in membrane structure. The viscoelasticity of membranes has been shown to be an unexpectedly rich area of study. In particular the observation of viscoelastic relaxation on microsecond time scales in these systems is completely unexpected. These are, as far as we know, the first observations of the macroscopic manifestations in membrane viscoelasticity of molecular relaxation. The time scales and their dependence upon temperature and molecular packing are strongly suggestive of possible biological significance.

However, to date few light scattering studies of surface or interfacial visco-elasticity have combined all of the above advantages. This has led to a reduction in the amount of information emerging from this work and may have hindered attempts to interpret the results of these studies.

The application of these methods to polymer films is still in its infancy. To date only the range of possibilities has been demonstrated. The potential for good and novel science would appear high, if the technique is applied to well-designed experiments addressing selected topics.

Polymer chains at the air–water interface can act as two-dimensional polymer solutions: the segments in contact with the fluid substrate are not in a pure state but appear as if in 'solution'. At least some of the film properties ($\pi - A$ isotherm) can be interpreted on the basis of scaling theories of two-dimensional semi-dilute polymer solutions. Further developments in this area may provide further insight into the other aspects of the film viscoelasticity discussed here.

In particular, it is hoped that the viscoelasticity may be related to the chain dynamics as in the case of three dimensions. For surface films of small amphiphiles, the surface viscoelasticity reflects the intramolecular conformational dynamics as well as the cooperative molecular translation. Such connections for polymer films would be most useful. If the light scattering reflects the phase transitions of the polymer films as for simpler systems, much information upon the former case could be derived.

REFERENCES

1. Kramer, L., *J. Chem. Phys.*, **55**, 2097 (1971).
2. Langevin, D., *J. Colloid Interf. Sci.*, **80**, 412 (1981).
3. Fan, C.-P., *J. Colloid Interf. Sci.*, **44**, 369 (1973).
4. Bouchiat, M. A., and Langevin, D., *J. Colloid Interf. Sci.*, **63**, 193 (1978).
5. Lucassen-Reynders, E. H., and Lucassen, J., *Adv. Colloid Interf. Sci.*, **2**, 347 (1969).
6. Lamb, H., *Hydrodynamics*, Dover, New York (1945).
7. Goodrich, F. C., *Proc. R. Soc. Lond.*, **A374**, 341 (1981).
8. Baus, M., *J. Chem. Phys.*, **76**, 2003 (1982).
9. Langevin, D., Meunier, J., and Chatenay, D., in K. L. Mittal and B. Lindman (eds.), *Surfactants in Solution*, Vol. 3, Plenum, New York (1984), pp. 1991–2014.
10. Earnshaw, J. C., and McGivern, R. C., *J. Colloid Interf. Sci.*, **123**, 36 (1988).
11. Chen, Y., Sano, M., Kawaguchi, M., Yu, H., and Zografi, G., *Langmuir*, **2**, 349 (1986).
12. Hård, S., and Neuman, R. D., *J. Colloid Interf. Sci.*, **83**, 315 (1981).
13. Hård, S., and Neuman, R. D., *J. Colloid Interf. Sci.*, **120**, 15 (1987).
14. Earnshaw, J. C., McGivern, R. C., Mclaughlin, A. C., and Winch, P. J., *Langmuir*, **6**, 649 (1990).
15. Earnshaw, J. C., McGivern, R. C., and Winch, P. J., *J. Phys. France*, **49**, 1271 (1988).
16. Ferry, J. D., *Viscoelastic Properties of Polymers*, John Wiley, New York (1980).
17. Earnshaw, J. C., and Winch, P. J., *J. Phys.: Condens. Matter*, **2**, 8499 (1990).
18. Crawford, G. E., and Earnshaw, J. C., *Biophys. J.*, **52**, 87 (1987).

19. Crawford, G. E., and Earnshaw, J. C., *Biophys. J.*, **55**, 1017 (1989).
20. Genz, A., and Holzwarth, J. F., *Colloid Polym. Sci.*, **263**, 484 (1985).
21. Robertson, R. N., *The Lively Membrane*, Cambridge University Press, Cambridge,, (1983), Ch. 1.
22. Papahadjopoulos, D., Jacobsen, K., Nir, S., and Isac, T., *Biochim. Biophys. Acta*, **311**, 330 (1973).
23. Crilly, J. F., and Earnshaw, J. C., *Biophys. J.*, **41**, 197 (1983).
24. However, cf. Earnshaw, J. C., McGivern, R. C., and Crawford, G. E. in S. E. Harding and A. J. Rowe (eds.), *Dynamic Properties of Biomolecular Assemblies*, Royal Society of Chemistry, Cambridge (1989), pp. 348–62.
25. Winch, P. J., and Earnshaw, J. C., *J. Phys. E: Sci. Instrum.*, **21**, 287 (1988).
26. Pallas, N. R., and Pethica, B. A., *Langmuir*, **1**, 509 (1985).
27. Winch, P. J., and Earnshaw, J. C., *J. Phys.: Condens. Matter*, **1**, 7187 (1989).
28. Moore, B., Knobler, C. M., Broseta, D., and Rondelez, F., *J. Chem. Soc., Faraday Trans. II*, **82**, 1753 (1986).
29. Middleton, S. R., and Pethica, B. A., *Faraday Symp., Chem. Soc.*, **16**, 109 (1981).
30. Crawford, G. E., and Earnshaw, J. C., *J. Phys. D*, **18**, 1029 (1985).
31. Earnshaw, J. C., and McGivern, R. C., *J. Colloid Interf. Sci.*, **131**, 278 (1989).
32. Crawford, G. E., and J. C. Earnshaw, *Biophys. J.*, **49**, 869 (1986).
33. Dilger, J. P., *Biochim. Biophys. Acta*, **645**, 357 (1981).
34. Kawaguchi, M., Sano, M., Chen, Y.-L., Zografi, G., and Yu, H., *Macromolecules*, **19**, 2606 (1986).
35. Kawaguchi, M., Sauer, B. B., and Yu, H., *Macromolecules*, **22**, 1735 (1989).
36. Chen, Y.-L., Kawaguchi, M., Yu, H., and Zografi, G., *Langmuir*, **3**, 31 (1987).
37. Sauer, B. B., and Yu, H., *Macromolecules*, **22**, 786 (1989).
38. Sauer, B. B., Yu, H., Tien, C.-F., and Hager, D. F., *Macromolecules*, **20**, 393 (1987).
39. Sauer, B. B., Kawaguchi, M., and Yu, H., *Macromolecules*, **20**, 2732 (1987).
40. Yoo, K.-H., and Yu, H., *Macromolecules*, **22**, 4019 (1989).

6

Characterization of Interfaces in Polymers and Composites Using Raman Spectroscopy

Robert J. Young

Manchester Materials Science Centre
UMIST/University of Manchester

1 INTRODUCTION

Over the past 15 years there has been an upsurge of interest in using Raman spectroscopy to characterize polymers. In particular, it has been found that the technique of Raman spectroscopy in conjunction with an optical microscope (the 'Raman microscope' or the 'Raman microprobe') has proved to be of particular use. The Raman effect is shown schematically in Figure 1 where a beam of light of frequency v_0 is scattered by a material. The majority of the radiation is scattered *elastically* at the same frequency (the Rayleigh scattering) but a small amount of the radiation is scattered *inelastically* at a different frequency $\pm \Delta v$ (the Raman scattering). It is this inelastically scattered light that carries information about molecular vibrations in the material and so is useful for the characterization of non-metallic materials.

The upsurge of interest in the application of Raman spectroscopy is due to at least two factors. Firstly, there have been significant developments in the area of instrumentation with lasers being more widely available and easier to use (e.g. being air-cooled rather than water-cooled) and the introduction of highly sensitive photon detectors such as charge-coupled device (CCD) cameras. Secondly, and probably most importantly, several new polymers have become available that are particularly amenable to characterization by Raman spectroscopy. High-performance fibres such as those obtained from liquid crystalline polymer solutions (e.g. Kevlar) have been developed over this period and have been shown to have particularly well-defined Raman spectra due to a combination of their highly oriented molecular structure and the aromatic nature of the molecular backbones. Conjugated polymers with interesting and unusual optical and electronic properties such as substituted polydiacetylenes have also been

Polymer Surfaces and Interfaces II
Edited by W. J. Feast, H. S. Munro and R. W. Richards
© 1993 John Wiley & Sons Ltd

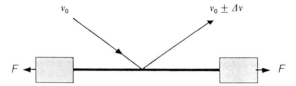

Figure 1. Schematic representation of the Raman effect. The situation shown is for a fibre deformation experiment where the laser beam of frequency v_0 is scattered from a fibre subjected to a force F

developed and their molecular structure again leads to their having particularly well-defined Raman spectra. This present chapter will be concerned with the use of Raman spectroscopy for the characterization of surfaces and interfaces in both polymers and polymer-based composites. Following a short review of the effect of deformation upon the Raman spectra of polymers, the instrumentation used in Raman spectroscopy will be described briefly and then the rest of the chapter will be concerned with the characterization of the surfaces and interfaces.

1.1   Deformation Studies Using Raman Spectroscopy

It has been found that the frequencies or wavenumbers of the Raman bands of many high-performance fibres shift on the application of stress to the fibre[1-16] and it is this observation that forms the basis of the present chapter. The phenomenon was first reported for the deformation of polydiacetylene single crystal fibres[1] and subsequently confirmed by others.[2-5] It is thought[2] that the behaviour is due to the macroscopic deformation being transformed directly into stressing of the covalent bonds along the polymer backbone and changes in the bond angles as discussed many years ago by Treloar.[17] The shift in the Raman band wavenumbers for polydiacetylenes are particularly large (-20 cm^{-1}/% strain) for the polydiacetylenes and their unique conjugated backbone structure leads to very strong and well-defined resonance Raman spectra[2] being obtained for these materials. Hence the phenomenon is particularly easy to follow in polydiacetylene single crystals.

Following this original work upon the polydiacetylenes it was found that stress- or strain-induced shifts in the position of Raman bands for several high-performance fibres could be obtained. This was first observed for aromatic polyamides such as Kevlar[6-8] and then for rigid-rod polymer fibres such as poly(p-phenylene benzobisthiazole) (PBT),[9] poly(p-phenylene benzoxazole) (PBO)[10] and poly(2,5(6)-benzoxazole) (ABPBO).[11] These fibres have well-defined Raman spectra due to the high degree of molecular orientation and also the possibility of resonance enhancement. They have very impressive physical properties such as a high value of Young's modulus[6–11] which means that during deformation the macroscopic deformation is translated to the direct stressing of the backbone bonds in the molecules which manifests as shifts in the bands positions in the spectra. Similar behaviour has also been found for gel-spun polyethylene fibres[12,13] and for other non-polymeric fibres such as those of carbon,[14] silicon carbide[15] and alumina.[16] The levels of shift measured for a variety of different high-performance fibres are listed in Table 1. Smaller frequency shifts (typically < -1 cm^{-1} up to yield) are found during the deformation of conventional polymers such as oriented polyesters[18] or polypropylene.[19] This is because deformation of these materials does not lead to significant direct stretching of the polymer backbone. Values of shift obtained for these conventional polymers are also listed in Table 1.

Following these original measurements of the stress- or strain-induced Raman band shifts it was soon realized that the phenomenon could be used to follow many different aspects of structure–property relationships in materials and it is the aim of this chapter to demonstrate how it can be used to characterize interfaces in polymers and polymer fibre reinforced composites.

Table 1. Peak frequency, Δv, and strain sensitivity, $S = d\Delta v/de$, of the Raman bands of various materials

Materials	Δv (cm^{-1})	$d\Delta v/de$ (cm^{-1}/%)^a	References
Polydiacetylene			
single crystals	2080	-20.0	1–5
Aramids	1613	-4.4	6–8
UHMW-PE	1060	~ -5	12,13
Rigid rods			
PBT	1477	-12.1	9
PBO	1280	-7.9	10
ABPBO	1555	-5.9	11
Non-polymeric fibres			
Carbon	1580	-14.6	14
Silicon carbide	1600	-6.6	15
Alumina–zirconia	460	-5.7	16
Conventional polymers			
PET	1615	< -1	18
Polypropylene	808	< -1	19

^{a} The data quoted are the highest values in the literature.

1.2 Instrumentation for Raman Spectroscopy

Most of the work in the literature concerning Raman spectroscopy upon polymers has been performed using dispersive spectrometers with photomultiplier detectors. Many of these systems incorporate microscopes for the alignment of the samples and collection of the spectra from regions as small as 2 μm in diameter. There have also recently been significant improvements in the hardware available for Raman spectroscopy. The introduction of array detectors has enabled a particular region of a spectrum to be acquired simultaneously rather than stepping through single points. This drastically reduces the time required to obtain a spectrum. A significant development in this area has been the introduction of charge-coupled device (CCD) cameras[20] with quantum efficiencies of between 20 and 40% in the region 450–900 nm and a dark noise of less than 10^{-3} electrons/pixel. A typical Raman microprobe system incorporating a CCD detector is shown schematically in Figure 2 and using such a system, the spectrum of Kevlar consisting of a single band in the region of 1580–1640 cm^{-1} can be obtained from areas of the order of 2 μm in diameter in a few seconds using a low-power 10 mW He–Ne laser.[8]

There have been recent reports[21,22] of the development of Fourier transform (FT) Raman systems which have certain advantages over conventional dispersive systems. The FT Raman systems are potentially highly efficient and

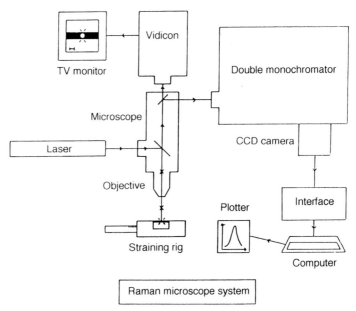

Figure 2. Schematic representation of a Raman microprobe (microscope) system incorporating a CCD detector

have a high calibration accuracy. One particular advantage for polymers is that the use of infra-red (IR) excitation wavelengths greatly reduces the problems of fluorescence. Unfortunately, however, the intensity of the Raman scattering is reduced by an order of magnitude by the use of IR radiation rather than excitation in the visible region because the scattering intensity depends on the fourth power of the excitation frequency, v_0^4. The choice between the use of either a dispersive or FT system depends upon the scattering characteristics of the material under investigation and the type of information being sought. Another significant advance is the development of the Raman microprobe into a true Raman microscope reported recently by Batchelder[23] whereby images of the Raman scattered light may be obtained. There is clearly plenty of scope for the use of these new systems for the analysis of surfaces and interfaces in polymers and there is no doubt that there will be many such reports in the years to come.

2 SKIN/CORE STRUCTURE IN KEVLAR FIBRES

2.1 Molecular Deformation Processes

Over the past 20 years there have been some very exciting developments of high-performance fibres with high levels of stiffness and strength. One of the first and best examples is fibres based on the aromatic polyamide poly(*p*-phenylene terephthalamide) (PPTA)[24,25] which have been commercialized by DuPont as Kevlar® and more recently by Akzo as Twaron®. The molecule of PPTA is shown below:

$$\left[-NH-\langle\bigcirc\rangle-CH_2-\langle\bigcirc\rangle-NH-CO-\langle\bigcirc\rangle-CO-\right]$$

The fibres are spun from concentrated liquid crystalline solutions of the polymer in solvents such as sulphuric acid. The as-spun fibres have good mechanical properties such as high stiffness and strength and the properties can be improved further by subjecting the fibres to heat treatment at elevated temperatures for short periods of time.

It has been found[26] that it is possible to use Raman spectroscopy to differentiate between the deformation of the surface skin and central core regions of Kevlar fibres and hence to characterize the structure of the fibre surfaces. Figure 3 shows the Raman spectrum for a Kevlar 49 fibre[24] between 1100 and 1700 cm^{-1} obtained using an He–Ne laser[26] in a Raman microprobe (Figure 2) and it can be seen that there are six well-defined peaks. Measurements of the absorption characteristics of films of the aromatic polyamide have shown that 90% of the visible light is absorbed by a depth of the order of 1 μm of the

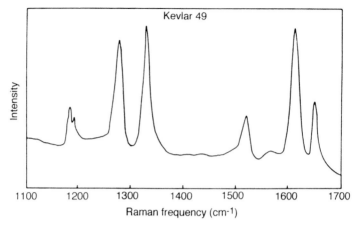

Figure 3. Raman spectrum obtained from a single Kevlar 49 aramid fibre. (After Ref. 27)

material[27] and so it is clear that the spectrum shown in Figure 3 comes from the skin regions of the 12 μm diameter Kevlar fibre.

It was described in Section 1.1 how Raman spectroscopy could be used to follow molecular deformation in a wide variety of high-performance fibres. An example of this is shown in Figure 4 for an experimental aromatic polyamide (aramid) fibre.[26] It is found that the position of the 1610 cm^{-1} Raman band shifts to a lower frequency under the application of a tensile stress as shown in Figure 4(a). The peak position of the band is plotted as a function of strain in Figure 4(b) and it can be seen that there is an approximately linear shift in peak position, Δv, with strain, e, up to failure at about 3% strain. The slope of the line is given by dΔv/de and it is normally found[6-8] that for different aramid fibres the slope is proportional to the fibre Young's modulus, E, i.e.

$$\frac{\mathrm{d}\Delta v}{\mathrm{d}e} \, \alpha E \qquad (1)$$

This is an indication that the Raman technique is directly probing molecular stretching and is consistent with the theoretical work of Northolt and co-workers.[28-30] They suggested that for the deformation of aramid fibres such as Kevlar the total strain is the sum of the strains due to two deformation processes, stretching and rotation, such that

$$e_{\text{total}} + e_{\text{stretch}} + e_{\text{rotation}} \qquad (2)$$

It has been shown[26] that the Raman technique follows only the crystal and molecular stretching and moreover this means that the change in the peak position, dΔv, is a measure of the molecular stress rather than the molecular

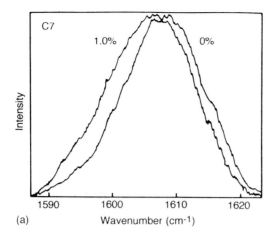

(a)

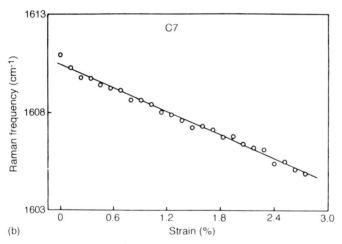

(b)

Figure 4. (a) Shift of the 1610 cm^{-1} Raman band with strain for an experimental aramid fibre (Table 2). (b) Dependence of the peak position upon strain. (After Ref. 26)

strain.[26] The rate of shift per unit strain in the higher modulus fibres is a reflection of the higher levels of molecular stress in such fibres.

2.2 Skin/Core Structure

The most direct method of characterizing the skin/core structure of aramid fibres is by the use of transmission electron microscopy (TEM). Figure 5 shows TEM sections obtained from an experimental aramid fibre with a Young's modulus of 90 GPa.[26] The fibre had been embedded in an epoxy mounting

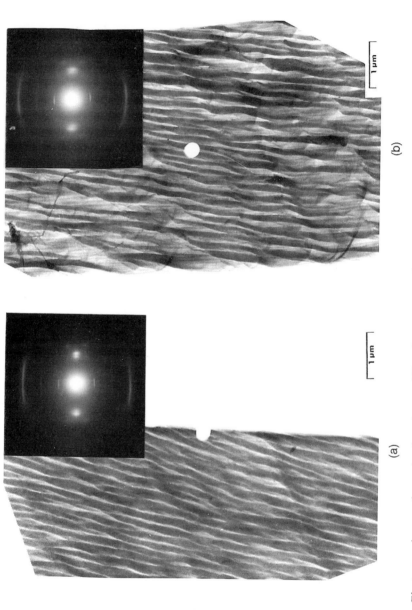

Figure 5. Electron micrographs and selected-area diffraction patterns (inset) obtained from ultra-microtomed sections of an experimental aramid fibre (B7, Table 2). The doubly exposed area indicates the position of the selected-area aperture. (a) Skin region. (b) Core region. (After Ref. 26)

(a)

(b)

1 μm

1 μm

resin and microtomed at room temperature using a diamond knife with the direction of the knife movement perpendicular to the fibre axis. The sections less than 100 nm thick were examined in a TEM operated at 100 kV, keeping the beam intensity at a low level to avoid beam damage.[26,27] Selected-area electron diffraction patterns (SADP) were obtained from different regions across the fibre section using a small (0.3 μm) selected-area aperture. Figure 5(a) shows a micrograph and SADP obtained from the surface skin region of the fibre and Figure 5(b) shows a similar section with an SADP from the fibre core. It is found that, for this experimental fibre, the angular azimuthal half-width of the arcs of the 110/200 equatorial peaks, $2\theta_h$,[26] is greater in the core regions than for the skin. This means that there is better molecular orientation in the surface skin region of the fibre than in the core.

The variation of $2\theta_h$ across the cross-section of another fibre (A7, Table 2) is shown in Figure 6(a) and it can be seen that $2\theta_h$ increases going from the surface of the fibre to the core. This experimental fibre was made by varying the processing conditions systematically to produce a large different in skin/core orientation. This generally leads to fibres with inferior mechanical properties such as lower strength and stiffness.[26] Commercial fibres such as Kevlar tend

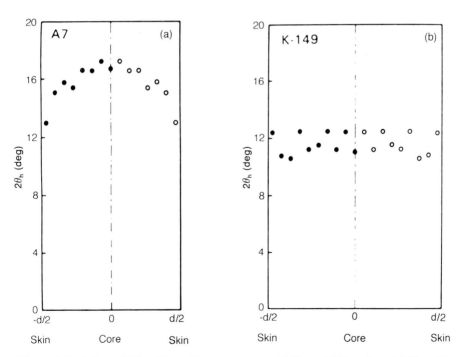

Figure 6. Variation of $2\theta_h$ with position across aramid fibres. (a) Experimental fibre A7 (Table 2). (b) Commercial fibre of the Kevlar 149 type. (After Ref. 26)

Table 2. Structure and property data for a series of experimental and commercial aramid fibres. (After Ref. 26)

Fibre	Diameter (μm)	$2\theta_h(°)$ skin	$2\theta_h(°)$ core	$\langle \sin^2 \phi \rangle$ skin	$\langle \sin^2 \phi \rangle$ core	Young's modulus (GPa)	Sonic modulus (GPa)	Fracture strength (GPa)	$d\Delta\nu/d\varepsilon$ (cm^{-1}/%)
A1	10.3	11.5	13.7	0.0212	0.0301	128	160	4.77	4.0
A7	10.2	12.6	17.5	0.0255	0.0489	94	111	3.27	4.0
B1	15.8	13.8	14.9	0.0305	0.0355	116	144	4.32	3.9
B7	14.0	12.3	17.2	0.0243	0.0473	90	112	3.09	3.6
C1	18.7	12.8	13.0	0.0263	0.0271	99	134	4.14	3.3
C3	19.1	14.3	15.9	0.0328	0.0404	79	112	3.07	2.8
C5	18.4	15.9	16.9	0.0404	0.0456	67	92	3.20	2.2
C7	19.1	16.6	18.2	0.0441	0.0529	67	98	3.05	2.1
D1	12.8	12.1	12.8	0.0235	0.0263	94	131	4.22	3.2
E1[a]	12.5	11.5	12.5	0.0212	0.0251	124	—	4.20	4.0
E2	11.9	11.9	12.9	0.0227	0.0267	129	156	4.13	4.0
F1[b]	12.4	11.4	11.7	0.0209	0.0216	161	143	3.30	5.2

[a] Similar to Kevlar 49.
[b] Similar to Kevlar 149.

to have more uniform properties with much less difference in skin/core orientation, as can be seen from Figure 6(b) which shows the variation of $2\theta_h$ with position for a Kevlar 149 fibre (Young's modulus 161 GPa).[26] It can be seen that the overall level of orientation is higher, reflected by the lower values of $2\theta_h$, consistent with a higher modulus, and that the value of $2\theta_h$ is approximately constant across the whole fibre section.

A number of measurements concerning the relationship between structure and mechanical properties for a series of commercial (Kevlar) and experimental aromatic polyamide fibres are listed in Table 2.[26] These include structural characterization of both the surface skin and core regions as well as the values of $d\Delta v/de$ determined using Raman spectroscopy. It is possible to make a large number of structure–property correlations from such data.[26] For instance, it has been shown[26] from the work of Northolt and coworkers[28–30] that the Young's modulus, E, of the fibres should be given by an equation of the form

$$\frac{1}{E} = \frac{1}{E_c} + \frac{\langle \sin^2 \phi \rangle}{2g} \tag{3}$$

where E_c is the Young's modulus of the PPTA crystal in the chain direction, $\langle \sin^2 \phi \rangle$ is an orientation parameter describing the average molecular orientation with respect to the fibre axis and g is the shear modulus in the chain direction. It is known[29] that if the azimuthal profile of the 110/200 reflection can be fitted to a Lorentz-IV distribution the orientation parameter $\langle \sin^2 \phi \rangle$ is related to the half-height width $2\theta_h$ approximately by

$$\langle \sin^2 \phi \rangle = 2.114 \sin^2 \theta_h \tag{4}$$

It has been found that a plot of the reciprocal of the sonic modulus of the fibres listed in Table 2 against the core value of $\langle \sin^2 \phi \rangle$ yields a straight line with an intercept corresponding to a modulus of the order of about 260 GPa, which is thought[26,30] to be close to the crystal modulus of PPTA, E_c. A similar analysis using the skin values of $\langle \sin^2 \phi \rangle$ does not give a straight line relationship[27] because the Young's modulus of the fibres is controlled by the mechanical properties of the core material which accounts for the vast bulk of the fibre. It has been shown,[26] however, that the presence of a highly ordered surface skin with a less-ordered core region can lead to fibres with inferior levels of strength.

Another aspect of the deformation of the surface skin regions is reflected in the dependence of $d\Delta v/de$ upon the fibre modulus, E, as given by equation (1). Figure 7 shows a plot of the data from Table 2 according to this equation. It can be seen that, as before,[7] most of the data points fall close to a straight line through the origin but that there are two points that lie significantly above the line. Inspection of Table 1 shows that these two experimental fibres have significant differences in molecular orientation between the surface skin and core regions reflected by differences in $2\theta_h$ of up to 5°. Since the He–Ne laser

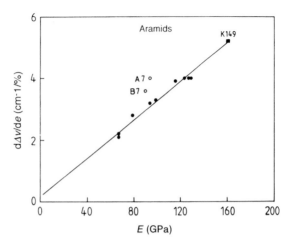

Figure 7. Dependence of $d\Delta\nu/de$ with Young's modulus E for the aramid fibres listed in Table 2. The line is a least squares fit of the solid data points. (After Ref. 26)

beam used in the Raman spectroscopy penetrates only the order of 1 μm deep into the aramid fibres[26,27] the spectra are obtained from only the surface skin regions. Hence for these two fibres the spectra come only from the highly oriented molecules in the surface regions which have higher Young's modulus values than for the core material. The values of $d\Delta\nu/de$ for these two fibres are higher than expected for a fibre of uniform structure and so the data points lie above the line in Figure 7. This demonstrates clearly the usefulness of using the technique of Raman spectroscopy to characterize the deformation of surfaces in aramid fibres.

3 INTERFACIAL MICROMECHANICS IN POLYMER FIBRE REINFORCED COMPOSITES

It was pointed out above that Raman spectroscopy can be used to follow deformation of high-modulus polymer fibres. In this section it will be demonstrated that the technique can be extended to investigate the micromechanics of deformation of the fibres within the matrix of an epoxy resin composite. This is an important development of the original single-fibre deformation studies[1-17] and has the potential of revolutionizing our understanding of the fundamental micromechanics of fibre reinforcement. The original investigations were undertaken by Young and coworkers[31-33] who studied model composites consisting of polydiacetylene single crystal fibres (Table 1) in a transparent epoxy resin matrix. They showed that strong well-defined Raman spectra could be obtained from local areas of the fibres using a focused laser beam with the epoxy resin matrix giving virtually no Raman scattering or fluorescence. Moreover, upon deformation of the epoxy resin matrix it was found[31,32] that the point-to-point variation of the strain within the fibre could be measured from the strain-induced shifts in the positions of the Raman bands for the single crystal fibres. For the first time this enabled the direct measurement of strain (and of stress in a linear elastic material) of a fibre within a composite.

The transfer of stress across the fibre–matrix interface is a fundamental problem in the understanding of composite micromechanics. It is the subject of a large number of theoretical investigations[34–36] and the Raman technique now allows this to be measured directly for real composite systems.

3.1 Rigid-rod Polymer Fibres in an Epoxy Resin Matrix

It has been demonstrated that for rigid-rod polymer fibres such as PBT, PBO and ABPBO (Table 3) the bands in their Raman spectra shift to a lower frequency under the action of tensile stress or strain.[9–11] This gives invaluable information upon the way in which rigid-rod polymer molecules deform in the fibres, leading to the fibres having very impressive mechanical properties.[9–11] This section will be concerned with the use of Raman spectroscopy to measure the point-to-point variation of stress within these three types of fibres in an epoxy resin composite.[37,38] Examples will be given for six fibres, i.e. both 'as-spun' (AS) and 'heat-treated' (HT) forms of PBT, PBO and ABPBO in a cold-setting epoxy resin matrix. Full details of the structure and mechanical properties of the fibres are given elsewhere.[9–11]

Single-fibre composite specimens were prepared consisting of individual fibres aligned in the axial direction in the centre of an epoxy resin tensile bar, cured

Table 3. Chain repeat units for various rigid-rod and stiff-chain polymers used in high-performance fibres[9–11]

Polymer	Chemical name	Repeat unit
PBT	Poly(*p*-phenylene benzobisthiazole)	
PBO	Poly(*p*-phenylene benzobisoxazole)	
ABPBO	Poly(2,5(6)-benzoxazole)	

at room temperature to avoid thermal stresses,[32] as described in detail elsewhere.[37,38] A resistance strain gauge was fixed to the side of the bar to allow measurement of the matrix strain. Raman spectra were obtained from the fibres within the resin bars using a Raman microprobe system (Figure 2) and either the 632.8 nm line of a 10 mW He–Ne laser or the 488 nm line of a 30 mW Ar$^+$ laser and the spectra were recorded using a liquid N_2 cooled CCD system.

The dependence of the positions of the Raman bands for the six different fibres upon strain has been reported elsewhere.[9–11] In general, it is found that for the HT fibres there is an approximately linear decrease in the frequency (Δv) of the band peak positions with strain, e, up to fracture, whereas for the AS fibres there is a linear decrease up to fibre yield with the frequency remaining approximately constant when the fibre is strained beyond yield. It is also found that for all of the fibres, $d\Delta v/de$ in the elastic region is approximately proportional to the fibre modulus (in the absence of skin/core effects). It has been shown[37] that this is an indication that the change in Raman frequency is proportional to the fibre *stress*. Hence, it is possible in the case of the rigid-rod fibres to determine directly the point-to-point variation of both stress and strain in the fibres in the resin during deformation.

The dependence of the fibre strain upon position along an HT ABPBO fibre in the single-fibre composite is shown in Figure 8. It can be seen that the

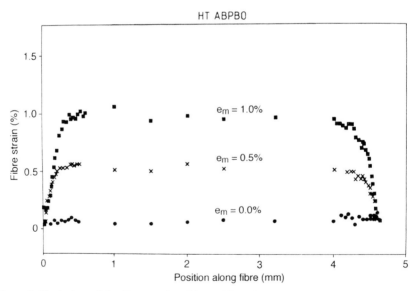

Figure 8. Variation of the fibre strain with position along the fibre for a single-fibre HT ABPBO epoxy resin composite. The results are given for three different levels of matrix strain e_m. (After Ref. 37)

behaviour is of the 'classical' Cox[34] shear-lag type, with the strain rising from zero at the fibre ends to equal the matrix strain in the central region of the fibre. In addition, it can be seen that in the central region of the fibre, the fibre strain equals the matrix strain. This type of behaviour is typical of all the rigid-rod polymer fibres in the composites at low strain ($<1\%$).[37,38] However, at higher strain levels the behaviour is quite different and can be summarized as follows:

(a) In the AS fibres, *fibre yielding* occurs for $e > 1\%$ and it is found that the fibre strain tends to lag behind the matrix strain at higher strains.
(b) In the HT fibres, *fibre fracture* tends to take place for $e > 1.3\%$, giving characteristic triangular distributions of strain in fibre fragments—very different from the shear-lag profiles.
(c) At high strain *matrix yielding* and *interfacial failure* take place, giving rise to trapezoidal distributions of strain along the fibres.

3.2 Interfacial Shear Stresses

Once the variation of the tensile stress and strain with position along the fibres is determined it is relatively easy to calculate the variation of the interfacial shear stress, τ_i, along the fibres.[37,38] In the work of Young and Ang[38] described here this was done by fitting the variation of stress, σ, with position, x, to a fifth-order polynomial and obtaining the derivative $d\sigma/de$. It can be readily shown[37,38] that

$$2\tau_i = r \frac{d\sigma}{dx} \tag{5}$$

where r is the fibre radius. The results of this analysis for an HT PBT fibre single-fibre composite deformed to 0.8% strain are shown in Figure 9. It can be seen that the variation of fibre stress with position is again classical.[34-36] In addition, the variation of τ_i with position is also presented and it can be seen that τ_i is a maximum at the end of the fibres and then falls to zero over the central regions of the fibres. This is exactly the form of behaviour expected from the Cox-type shear-lag analysis.[34-36] It appears that the fibre-matrix adhesion for the HT PBO fibres was not as good as for the HT PBT ones, as can be seen from Figure 10. It can be seen that there is a trapezoidal distribution of tensile stress along the fibres and that τ_i peaks at some distance away from the end of the fibres, due probably to breakdown of the fibre–matrix interface. If it were due to matrix yielding, similar behaviour would have been expected for the PBT fibre composite in Figure 9.

Young and Ang[38] reported that they performed a large number of experiments for composites containing the six types of fibres and deformed to different

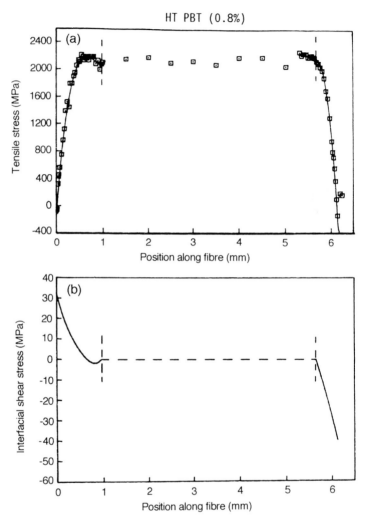

Figure 9. (a) Variation of tensile stress with position along an HT PBT fibre in a single-fibre composite subjected to an overall strain of 0.8%. (b) Variation of interfacial shear stress with position along the fibre in (a) derived using equation (5). (After Ref. 38)

levels of overall strain. In each case the distribution of fibre stress along the fibres was measured and τ_i determined. The maximum values of τ_i for the six different fibres are listed in Table 4. As far as is known, none of the fibres were given any special surface treatments. It can be seen from the table that τ_i is generally higher for the AS fibres than for the HT fibres. The low value of τ_i for HT PBO is consistent with the behaviour shown in Figure 10.

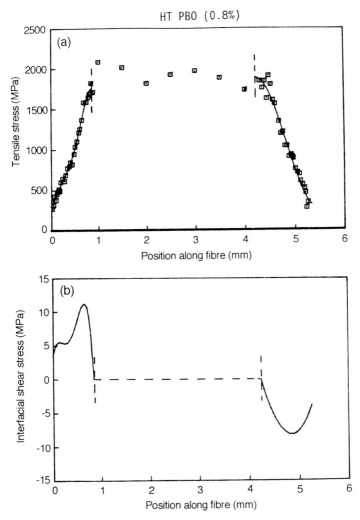

Figure 10. (a) Variation of tensile stress with position along an HT PBO fibre in a single-fibre composite subjected to an overall strain of 0.8%. (b) Variation of interfacial shear stress with position along the fibre in (a) derived using equation (5). (After Ref. 38)

Table 4. Maximum interfacial shear stresses ($\tau_i \pm 10\%$) for the different fibres. (After Ref. 38).

Fibre	AS PBT	HT PBT	AS PBO	HT PBO	AS ABPBO	HT ABPBO
τ_i(MPa)	40	36	50	15	54	23

3.3 General Applications

In conclusion it can be seen that it has been demonstrated that Raman spectroscopy is a powerful method of following the interfacial micromechanics of the deformation of rigid-rod polymer fibres within composites. It is possible to do this quantitatively and determine such fundamental parameters such as the interfacial shear stress, τ_i. Although the technique has been demonstrated only for model single-fibre composites containing rigid-rod fibres it is clear that it is of considerably more general use[39,40] and can be applied to a very wide range of high-performance fibres in a large number of matrix systems.

4 OPTICAL STRAIN-SENSITIVE SURFACE COATINGS

All of the investigations described above have been concerned with high-modulus polymer fibres. Their structure consists of highly oriented molecules aligned parallel to the fibre axis. They therefore have highly anisotropic mechanical properties with very high levels of Young's modulus parallel to the fibre axis and Young's modulus values of about two orders of magnitude lower in transverse directions. Although these materials have very impressive mechanical properties, especially when deformed parallel to the fibre axis, they only have limited applications and the vast majority of polymers are used in the unoriented state and have isotropic mechanical properties. It is therefore of interest to see to what extent Raman spectroscopy can be used to characterize interfaces in isotropic polymers. Early attempts to do this for unoriented polypropylene were disappointing[19] because deformation of this material does not lead to significant levels of molecular stretching. Shifts of little more than $-1\,\mathrm{cm}^{-1}$ were obtained for deformation up to yield.

Over recent years a completely different approach has been adopted by Hu, Day, Stanford and Young[41-44] who have shown that, through the synthesis of specially designed copolymers, it is possible to prepare isotropic polymers for which the deformation can be followed using Raman spectroscopy. They have demonstrated that such materials can be used for the study of polymer surface and interface deformation[43-44] and their work in this area is reviewed below.

4.1 Urethane–Diacetylene Copolymers

The approach that has been adopted[41-44] is to prepare a series of urethane diacetylene copolymers in which polydiacetylene units are incorporated into segmented copolyurethanes. It was pointed out earlier that large strain-induced Raman band shifts can be obtained during the deformation of polydiacetylene single crystal fibres.[1-5] In fact, it is found that the largest shift measured so far ($-20\,\mathrm{cm}^{-1}$/% strain) is for the $C{\equiv}C$ triple-bond stretching band of poly-diacetylenes (Table 1).

Polyurethanes constitute a versatile class of materials ranging from soft elastomers to glassy resins and are readily produced by a variety of processes into many different forms. Fibres, films, bulk sheets and surface coatings can all be formed either from solution or in bulk. Segmented copolyurethanes, because of their phase-separated structures, are particularly attractive since they enable the combination of disparate polymer properties to be obtained within a single material. Polydiacetylene single crystals are formed by the rapid solid-state polymerization[4] of substituted diacetylene monomer crystals on the application of heat or radiation. It is possible to induce similar solid-state reactions known as 'cross-polymerization' in the diacetylene groups of repeat units in certain copolyurethanes and copolyesters.[45-47] Cross-polymerization within the crystalline diacetylene regions produces a network structure in which the chains connecting the polydiacetylenes are analogous to the substituent side-groups in the polydiacetylene single crystals.[4] In this way, linear copolyurethanes containing phase-separated, diacetylene-containing domains, can be crosslinked *in situ* and transformed into insoluble and infusible materials.[48-50] These previous studies were concerned only with elastomeric materials formed via a two-stage solution process[48-50] at only one composition. The more recent studies[42-44] have been concerned with the development of more rigid copolymers formed by a relatively simple one-shot bulk polymerization route.

A reaction scheme that can be used to prepare the copolyurethanes is shown in Scheme I. The reactants used are 4,4′-methylenediphenylene diisocyanate (MDI), 2,4-hexadiyne-1,6-diol (HDD) and a polypropylene glycol (PPG400). It is necessary that all reactions are carried out using stoichiometric equivalent

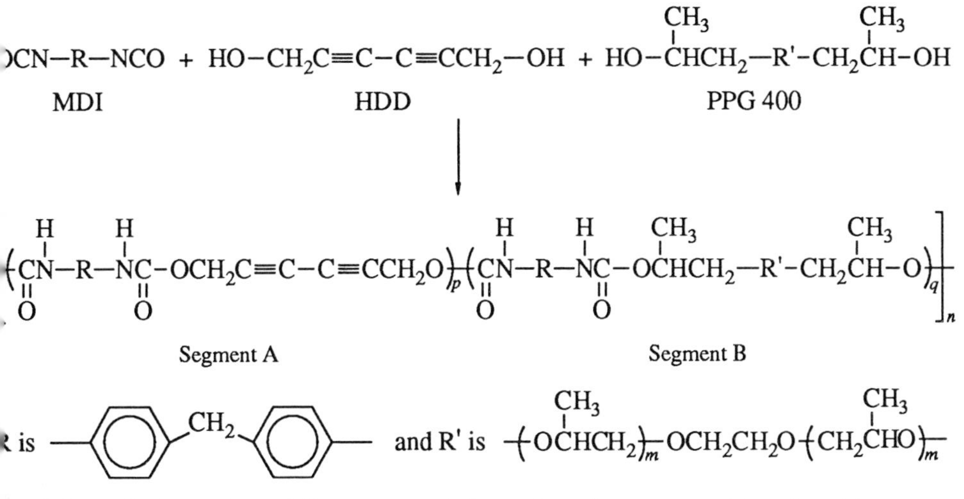

Scheme I. Formation of linear segmented copolyurethanes comprising diacetylene–urethane hard segments, A, and polyether–urethane segments, B

amounts of isocyanate and total hydroxyl groups, although it is possible to vary the structure and consequent properties of the materials by varying the relative proportions of HDD and PPG400. The exact details of the reaction conditions are given elsewhere[42] and linear polymers can be produced with molar masses of the order of 10 000 g/mol. These materials are in the form of off-white transparent/translucent sheets which need to be stored in a refrigerator in the dark as they undergo cross-polymerization in the laboratory at room temperature. The linear segmented urethane copolymers are soluble in a variety of solvents and can therefore be processed into a variety of forms such as surface coatings.[43]

It can be seen from the reaction in Scheme I that the structure of the material consists of diacetylene-urethane hard segments, A, and polyether–urethane segments, B. The development of this alternating segmented structure results in phase separation due to incompatibility of the chemically distinct segments (A and B). Thermodynamic incompatibility depends primarily on the interaction parameter between the diacetylene- and polyether-based segments (determined by their intrinsic solubility parameters) and their sequence lengths (degrees of polymerization). The development of hydrogen bonding and potential crystallinity of the hard segments further enhance the driving force for phase separation. The linear copolyurethanes thus form as essentially a two-phase morphology consisting of rigid, highly hydrogen-bonded hard segments domains (with a distribution of sizes) dispersed in a ductile, polyether–urethane phase. The overall degree of phase separation also depends on the nature of the interphase regions and on the mutual contamination of the phases by alternative segments. The formation of linear diacetylene-containing copolyurethanes provides the distinct advantage, in subsequent applications, of enabling the copolymers to be processed from solution. During or after removal of the solvent, the phase-separated copolymers can be rapidly cross-linked *in situ*, using heat or radiation, either of which causes cross-polymerization of diacetylene units within the hard segment domains. The basic chemical reaction of the diacetylene unit is shown simplistically in Scheme II in the terms of a single full-extended chain.

In practice, however, the solid-state topochemical reaction involves many such chains packed within the hard segment domains, and the resulting cross-polymerization occurs three dimensionally as depicted in Figure 11. The diacetylene–urethane hard segments are assumed to be crystalline and have fully extended conformations in which the HDD unit is all-*trans* and the chains are staggered so that adjacent chains are linked by straight $C=O\cdots H-N$ hydrogen bonds in both directions perpendicular to the urethane chain axes. It has been found[42] that these hard segment domains are organized in the form of spherulitic entities which are seen to be of the order of 1 μm in diameter using transmission electron microscopy.

The idealized structure in Figure 11, however, is unlikely to be totally

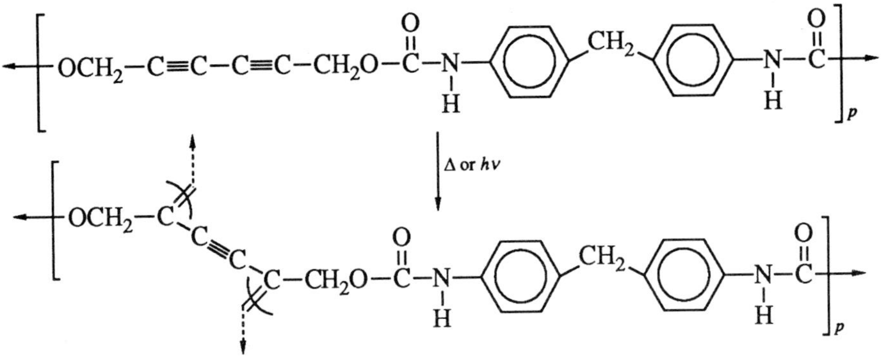

$$\left[\text{OCH}_2-\text{C}\equiv\text{C}-\text{C}\equiv\text{C}-\text{CH}_2\text{O}-\overset{\overset{\text{O}}{\|}}{\text{C}}-\underset{\underset{\text{H}}{|}}{\text{N}}-\hspace{-0.3em}\left\langle\bigcirc\right\rangle\hspace{-0.3em}-\text{CH}_2-\hspace{-0.3em}\left\langle\bigcirc\right\rangle\hspace{-0.3em}-\underset{\underset{\text{H}}{|}}{\text{N}}-\overset{\overset{\text{O}}{\|}}{\text{C}} \right]_p$$

$$\downarrow \Delta \text{ or } h\nu$$

$$\left[\text{OCH}_2-\text{C}\cdots\text{C}\equiv\text{C}\cdots\text{C}-\text{CH}_2\text{O}-\overset{\overset{\text{O}}{\|}}{\text{C}}-\underset{\underset{\text{H}}{|}}{\text{N}}-\hspace{-0.3em}\left\langle\bigcirc\right\rangle\hspace{-0.3em}-\text{CH}_2-\hspace{-0.3em}\left\langle\bigcirc\right\rangle\hspace{-0.3em}-\underset{\underset{\text{H}}{|}}{\text{N}}-\overset{\overset{\text{O}}{\|}}{\text{C}} \right]_p$$

Scheme II. The reaction of the diacetylene unit in a rigid diacetylene–urethane hard segment to produce a fully conjugated polydiacetylene unit

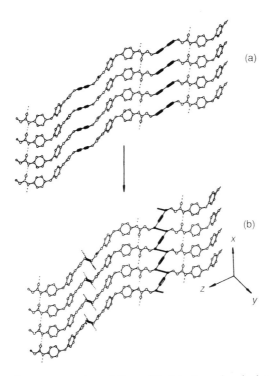

Figure 11. Schematic representation of the solid-state topochemical polymerization of the diacetylene–urethane hard segments. (a) Linear segmented block copolymer. (b) Cross-polymerized material. (After Ref. 42)

representative of the overall structure actually obtained for the hard segment domains, although regions of such a three-dimensional order must exist within domains dispersed throughout the copolyurethanes in order to achieve overall cross-polymerization. The formation of fully conjugated polydiacetylene (PDA) chains within the phase-separated copolyurethanes produces dramatic colour changes (white → red → deep purple) and transforms the copolymers into completely insoluble and infusible materials. The extent of cross-polymerization that is achieved depends upon a number of factors such as the time and temperature of heating, the concentration of hard segment domains and the degree of molecular order within the domains.[42]

The relationship between chemical composition, structure and properties for the copolymers has been described in detail elsewhere.[42,43] In general it is found that the glass transition temperature, T_g, and Young's modulus, E, increase with hard segment content and heat treatment temperature. It was found that the material with the optimum composition and properties had a value of T_g of about 80°C and a Young's modulus (isotropic) of about 1.7 GPa, both of which are typical of a conventional glassy polymer.

4.2 Raman Deformation Studies

A Raman spectrum for a sample of the cross-polymerized copolyurethane is shown in Figure 12 and it can be seen that it has four main scattering bands.[42,43] The spectrum is remarkably similar to that obtained from a polydiacetylene single crystal fibre.[4] For the fibres, a strong spectrum is only obtained when

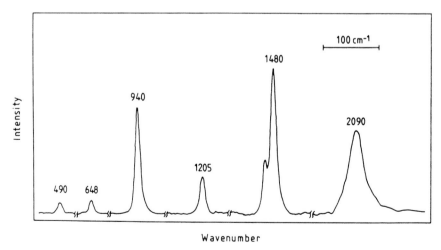

Figure 12. Full Raman spectrum for the cross-polymerized diacetylene–urethane copolymer. (After Refs. 42 and 43)

the fibre axis is parallel to the direction of polarization of the laser beam whereas it was found that the spectrum from the copolymer was identical for all orientations of the sample.[42,43] This shows clearly the isotropic nature of the copolymer. It was found[42,43] that the intensities of the bands in the Raman spectrum varied with both the hard segment content and heat treatment temperature and, moreover, it was demonstrated that this variation could be used to follow the cross-polymerization reaction. In particular the C≡C triple-bond stretching band at about 2090 cm^{-1} is not present before cross-polymerization and is indicative of the formation of the polydiacetylene chains in the structure.

Figure 13 shows the position of the C≡C Raman band for one of the copolymers in the undeformed and deformed states.[43] Upon deformation there is a pronounced shift to lower frequency and slight broadening of the Raman band. This shift is a clear indication of stress transfer from the polyether–urethane matrix to the diacetylene–urethane hard segments and shows that this stress is translated into direct deformation of the polydiacetylene chains. Broadening of the bands suggests that a distribution of stresses is developed within the non-uniform hard segments domains during deformation.

The dependence of the band position upon tensile strain is shown in Figure 14. It can be seen that it is approximately linear up to about 1% strain with a slope, $d\Delta v/de$, of the order of about -5 cm^{-1}/% strain and is also completely reversible on both unloading and reloading. The slope of the line in Figure 14 is significantly lower than that for polydiacetylene single crystal fibres (Table 1) but is comparable to that of high-performance fibres such as those based upon aromatic polyamides or rigid-rod polymers.

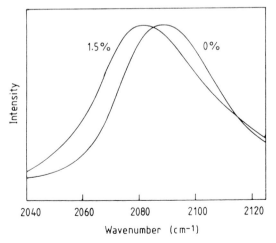

Figure 13. Shift of the 2090 cm^{-1} Raman band with strain for the cross-polymerized urethane–diacetylene copolymer. (After Ref. 43)

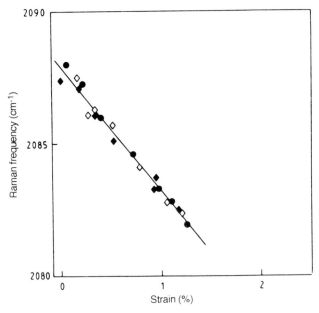

Figure 14. Dependence of the peak position of the 2090 cm^{-1} Raman band upon strain
($\blacklozenge$, first loading; $\bullet$, unloading; $\diamond$, reloading). (After Ref. 43)

It has been found that the relationship shown in Figure 14 can be used to measure strain in a wide variety of situations. It is known that polydiacetylenes absorb visible light very strongly and so for the bulk copolyurethane the spectrum is only obtained from material in the surface regions. Hence any strain measurements will be only for surface material. Moreover, since it is possible to focus the laser beam in the spectrometer to a spot of the order of 2 μm in diameter it is possible to obtain considerable spatial resolution.

Various examples of using these materials for surface-strain mapping have been presented in a recent publication.[44] The determination of stress concentrations around defects such as hole or notches in a deformed plate of the copolyurethane is shown in Figure 15. A circular hole and a notch of predetermined dimensions were accurately machined into a 3 mm thick specimen of the copolymer. The specimen was deformed in tension in the Raman spectrometer and the change in position of the C≡C stretching band at 2090 cm^{-1} with copolymer strain (measured remotely with a resistance strain gauge) was determined at different positions around the defects, as shown schematically in the insert in Figure 15. The slope (dΔv/de) of each line, relative to that for the remote applied deformation at (open $\bigcirc$), is proportional to the stress concentration at each position. For the hole, the highest stress concentration, as expected, is at the equator (solid $\bullet$) and is essentially zero at the pole

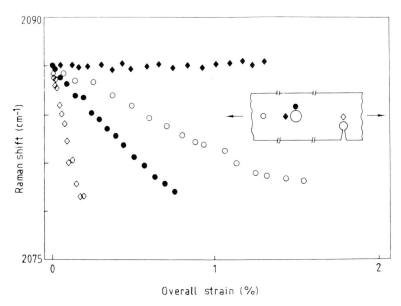

Figure 15. The effects of stress concentrations on the peak position of the C≡C Raman band for a 3 mm thick cross-polymerized diacetylene–urethane plate deformed in tension. The data were obtained from spectra obtained at the different position indicated. (After Ref 44)

(solid ◆). For the notch, the stress concentration (open ◇) increases sharply depending on the notch tip radius and distance from the tip. The results obtained for the various defect geometries[44] show the stress concentration values measured by Raman spectroscopy to be very similar to those determined from conventional stress analyses.[51]

The good solubility and adhesive characteristics of the as-prepared, di-acetylene-containing copolyurethanes make them attractive materials for subsequent use as surface coatings that can be applied with controlled thickness to a variety of substrates. Subsequent cross-polymerization, *in situ*, would then convert the coatings into cross-linked materials with strain-sensitive properties that can be determined quantitatively, in conjunction with the substrate, using Raman spectroscopy. To illustrate this use[44] a solution of the copolyurethane was applied as 0.05 mm coatings to the following substrates: (a) a sheet of highly cross-linked (non-diacetylene containing) polyurethane resin, (b) an inorganic glass filament ($\approx 25 \, \mu$m diameter) and (c) a sheet of aluminium. The solvent was removed by evaporation and the coatings were thermally treated at 100°C for 40 hours. The coated substrate specimens were deformed in tension in a Raman spectrometer and the shift in the position (Δv) of the C≡C triple-bond stretching band was monitored as a function of the overall specimen strain, e.

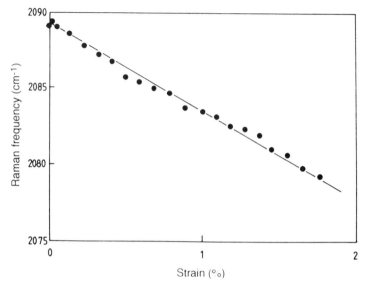

Figure 16. Dependence of the position of the C≡C Raman band upon tensile strain for the cross-polymerized diacetylene–urethene coated on the surface of a glass filament. (After Ref. 44)

Excellent linearity between Δv and e was obtained in each case as shown, for example, in Figure 16. The strain sensitivities of the three coated substrates determined from the slopes of the plots are given in Table 5 and the value of -5 cm^{-1}/% strain for dΔv/de are almost identical to that of the bulk sheet material. Clearly, these results demonstrate that the copolymers can be used as coating to monitor accurately the deformation of a substrate, which is particularly useful if the substrate is of complex geometrical shape or is not readily accessible for direct measurements. As such, the polydiacetylene-containing copolyurethanes are shown to behave as optical strain gauges.

Table 5. Strain sensitivities of the Raman frequencies of the C≡C triple-bond stretching band for the diacetylene–urethane copolymer and for the copolymer coated on different substrates. (After Ref. 44)

Substrate	dΔv/de (cm^{-1}/%)
Pure copolymer	5.3 ± 0.4
Polyurethane	5.5 ± 0.4
Glass fibre	5.7 ± 0.4
Aluminium	5.5 ± 0.4

5 CONCLUSIONS

It has been demonstrated that Raman spectroscopy is an excellent method of characterizing interfaces in polymers and composites. It has been shown that high-modulus polymer fibres have extremely well-defined Raman spectra and that the bands in the spectra shift to a lower frequency under the action of stress or strain. This provides a unique insight into the effect of macroscopic deformation upon the molecules in the structure and allows the relationship between structure and mechanical properties to be studied in detail. In particular, it has been demonstrated that it is possible to identify skin/core differences in Kevlar fibres and to follow, in detail, deformation of the fibre surface skin.

It has been shown that an extension of the ability to study molecular deformation in the fibres is to use Raman spectroscopy to follow the deformation micromechanics of high-performance fibres in composites. In particular, it has been shown that it is possible to measure the interfacial shear strength for individual fibres in an epoxy resin matrix. This has exciting implications for the study of composite micromechanics in a wide variety of systems.

Finally, recent developments in the preparation of optical strain-sensitive coatings have been reviewed. It has been shown that certain glassy diacetylene–urethane copolymers can be prepared with isotropic mechanical properties that have well-defined Raman spectra in which the bands shift on the application of strain. These materials have considerable promise as optical strain gauges and examples have been presented of how they can be used to measure stress concentrations at surfaces in a variety of different deformation geometries.

In conclusion, it is clear that the use of Raman spectroscopy for surface and interface characterization will increase significantly in the years to come.

REFERENCES

1. Mitra, V. K., Risen, W. M., and Baughman, R. H., *J. Chem. Phys.*, **66**, 2731 (1977).
2. Batchelder, D. N., and Bloor, D., *J. Polym., Sci.: Polym. Phys. Edn*, **17**, 569 (1979).
3. Galiotis, C., Young, R. J., and Batchelder, D. N., *J. Polym. Sci.: Polym. Phys. Edn*, **21**, 2483 (1983).
4. Young, R. J., in I. M. Ward (ed.), *Developments in Oriented Polymers*, Vol. 2, Elsevier, London (1987), p. 1.
5. Wu, G., Tashiro, K., and Kobayashi, M., *Macromolecules*, **22**, 188 (1989).
6. Galiotis, C., Robinson, I. M., Young, R. J., Smith, B. E. J., and Batchelder, D. N., *Polym. Commun.*, **26**, 354 (1985).
7. van der Zwaag, S., Northolt, M. G., Young, R. J., Galiotis, C., Robinson, I. M., and Batchelder, D. N., *Polym. Commun.*, **28**, 276 (1987).
8. Young, R. J., Lu, D., and Day, R. J., *Polym. Int.*, **24**, 71 (1991).
9. Day, R. J., Robinson, I. M., Zakikhani, M., and Young, R. J., *Polymer*, **28**, 1883 (1987).
10. Young, R. J., Day, R. J., and Zakikhani, M., *J. Mater. Sci.*, **25**, 127 (1990).
11. Young, R. J., and Ang, P. P., *Polymer*, **33**, 975 (1992).

12. Prasad, K., and Grubb, D. T., *J. Polym. Sci., Polym. Phys. Edn*, **27**, 381 (1989).
13. Kip, B. J., van Eijk, M. C. P., and Meier, R. J., *J. Polym. Sci.*, **B29**, 99 (1991).
14. Robinson, I. M., Zakikhani, M., Day, R. J., Young, R. J., and Galiotis, C., *J. Mater. Sci. Lett.*, **6**, 1212 (1987).
15. Day, R. J., Piddock, V., Taylor, R., Young, R. J., and Zakikhani, M., *J. Mater. Sci.*, **24**, 2898 (1989).
16. Yang, X., Hu, X., Day, R. J., and Young, R. J., *J. Mater. Sci.*, **27**, 1409 (1992).
17. Treloar, L. R. G., *Polymer*, **1**, 95 (1960).
18. Fina, L. J., Bower, D. I., and Ward, I. M., *Polymer*, **29**, 2146 (1988).
19. Evans, R. A., and Hallam, H. E., *Polym. Lett.*, **17**, 839 (1976).
20. Batchelder, D. N., *Euro. Spectrosc. News*, **80**, 28 (1988).
21. Zimba, C. G., Hallmark, V. M., Swalen, J. D., and Radbolt, J. F., *Appl. Spectrosc.*, **41**, 721 (1987).
22. Purcell, F. J., *Spectrosc. Int.*, **1**, 33 (1990); *Spectroscopy*, **4**, 24 (1989).
23. Batchelder, D. N., University of Leeds, personal communication.
24. Schaefgen, J. R., in A. E. Zachariades and R. S. Porter (eds.), *Strength and Stiffness of Polymers*, Marcel Dekker, New York, pp. 339.
25. Yang, H. H., *Aromatic High-Strength Fibres*, John Wiley, New York (1989).
26. Young, R. J., Lu, D., Day, R. J., Knoff, W. F., and Davis, H. A., *J. Mater. Sci.*, **27**, 5431 (1992).
27. Lu, D., Structure/property relationships in aromatic polymide fibres, PhD thesis, Victoria University of Manchester (1991).
28. Northolt, M. G., and van Aartsen, J. J., *J. Polym. Sci., Polym. Symp.*, **58**, 283 (1977).
29. Northolt, M. G., *Polymer*, **21**, 1199 (1980).
30. Northolt, M. G., and van de Hout, R., *Polymer*, **26**, 310 (1985).
31. Galiotis, C., Young, R. J., Young, P. H. J., and Batchelder, D. N., *J. Mater. Sci.*, **19**, 3680 (1984).
32. Robinson, I. M., Young, R. J., Galiotis, C., and Batchelder, D. N., *J. Mater. Sci.*, **22**, 3642 (1987).
33. Robinson, I. M., Galiotis, C., Batchelder, D. N., and Young, R. J., *J. Mater. Sci.*, **26**, 2293 (1991).
34. Cox, H. L., *Br. J. Appl. Phys.*, **3** 72 (1952).
35. Termonia, Y., *J. Mater. Sci.*, **22**, 1733 (1987).
36. Hull, D., *An Introduction to Composite Materials*, Cambridge University Press (1981).
37. Ang, P. P., Deformation micromechanics of high modulus polymer fibres and composites, PhD thesis, Victoria University of Manchester (1991).
38. Young, R. J., and Ang, P. P., in I. Verpoest and F. R. Jones (eds.), *Interfacial Phenomena in Composite Materials '91*, Butterworth-Heinemann, Oxford (1991), p. 49.
39. Guild, F. J., Galiotis, C., Jahankhani, H., and Vlattas, C., in I. Verpoest and F. R. Jones (eds.), *Interfacial Phenomena in Composite Materials '91*, Butterworth-Heinemann, Oxford (1991), p. 26.
40. Galiotis, C., and Jahankhani, H., in J. Füller, G. Grüninger, K. Schulte, A. R. Bunsell and A. Massiah (eds.), *Developments in the Science and Technology of Composite Materials*, Elsevier Applied Science (1990), p. 679.
41. Stanford, J. L., Young, R. J., and Day, R. J., *Polymer*, **32**, 1713 (1991).
42. Hu, X., Stanford, J. L., Day, R. J., and Young, R. J., *Macromolecules* (1992), **25**, 672 (1992).
43. Hu, X., Stanford, J. L., Day, R. J., and Young, R. J., *Macromolecules* (1992), **25**, 684 (1992).
44. Hu, X., Day, R. J., Stanford, J. L., and Young, R. J., *J. Mater. Sci.*, **27**, 5958 (1992).

45. Wegner, G., *Makromol. Chem.*, **134**, 219 (1970).
46. Day, D., and Lando, J. B., *J. Polym. Sci., Polym. Lett. Edn*, **19**, 227 (1981).
47. Patil, A. O., Desphande, D. D., Talwar, S. S., and Biswas, A. B., *J. Polym. Sci., Polym. Chem. Edn*, **19**, 1155 (1981).
48. Rubner, M. F., *Macromolecules*, **19**, 2114 (1986).
49. Liang, R. S., and Reiser, A. J., *J. Polym. Sci., Polym. Chem. Edn*, **25**, 451 (1987).
50. Nallicheri, R. A., and Rubner, M. F., *Macromolecules*, **24**, 517 (1991).
51. Williams, J. G., *Stress Analysis of Polymers*, Longman, London (1973).

7

Surface Modification and Analysis of Ultra-high-modulus Polyethylene Fibres for Composites

Graeme A. George

Department of Chemistry,
The University of Queensland

1 INTRODUCTION

Fibre-reinforced composite materials consist of high-strength, high-stiffness fibres of diameter 10–30 μm embedded in a thermosetting or thermoplastic matrix.[1] The volume fraction of the fibre is generally high (e.g. 0.7) in order to achieve the high stiffness and strength of the fibre in the composite. This results in a specific surface area of the fibre from 0.1 to 1.0 $m^2 g^{-1}$ so the interface between the fibre and the matrix is a major component of the material.

While early composite materials engineers paid little attention to the nature of the interface, arguing that the mechanical properties of the fibre dominated the tensile and compressive properties of the composite, it soon became apparent that the achievement of high fracture toughness and interlaminar shear strength required control of the interfacial properties. For a brittle fibre such as graphite, the optimization of strength requires efficient stress transfer between the fibre and the matrix while high fracture toughness requires controlled debonding of the fibre at the advancing crack tip.[2] These opposing requirements mean that great attention has now been paid to understanding and, if possible, controlling the chemistry at the interface between the matrix and the fibre.[3] This has resulted in a focus on micromechanical properties of a fibre embedded in a composite matrix, on the one hand, and the surface chemistry and analysis of the fibre, on the other. Recently this has been widened by recognizing that the region up to 1 μm from the fibre surface into the thermoset matrix will have a different composition because of the preferential reactivity of components in the resin as well as the presence of sizing agents.[4] This has been termed the 'interphase' and various proposals have been advanced for its structure and significance.[4,5] It has been noted that there may

Polymer Surfaces and Interfaces II
Edited by W. J. Feast, H. S. Munro and R. W. Richards
© 1993 John Wiley & Sons Ltd

Table 1. Typical mechanical and surface properties of some high-modulus fibres and matrix materials for composites[6-9,33,64-65]

Material	Youngs modulus E(GPa)	Tensile strength σ_f(GPa)	Elongation at fail ε_f(%)	Density ρ(g/cm³)	Critical surface tension γ_c(mN/m)	Fibre knot strenth (GPa)
Carbon						
High Modulus	400	2.0	0.6	1.8	42	0
High Strength	225	3.0	1.3			
E glass	69	2.4	4.0	2.5	47	0
Aramid (Kevlar 49)	125	2.6	2.4	1.44	43.7-46.4	0.6-0.8
UHMPE (Spectra 1000) (Dyneema)	170	3.0	2.7	0.97	31-36	1.1-1.7
Epoxy (DGEBA-amine)	3.0-3.6	0.062-0.082	3.0-5.0	1.2	43-47	—
Unsaturated polyester (Styrene cross-linked)	3.5-4.0	0.048-0.075	1.7-4.0	1.2	36	—

also be an interphase extending into the fibre as well, since carbon fibres have outer layers which differ in the extent of graphitization from the bulk.[6]

Over the past fifteen years high modulus organic fibres have become commercially available.[7] The first class were the aromatic polyamides such as Kevlar 49® aramid. These are rigid-rod molecules which form liquid crystals in the spinning solution. More recently, extended-chain polyethylene fibres have become available under the trade names Dyneema® and Spectra®.[8] In these fibres, ultra-orientation of the carbon backbone is achieved in an attempt to approach the theoretical strength of the totally extended chain (33 GPa). As can be seen from Table 1, this is far from realized in actual fibres, but the strength and stiffness approaches that of the carbon and E glass fibres. A major difference between the fibres is the failure mode. Glass and carbon fail in tension with a clean fracture surface while the aramids and extended-chain polyethylene fail by longitudinal splitting. This reflects the ultra-orientation and the anisotropic nature of organic fibres. The large surface area created at failure results in the superior impact and ballistic performance of these fibres.[8]

Micromechanical studies of high-modulus fibres in polymer matrices have been most extensive for brittle fibres,[3-5] but there are few clear indications as to the interfacial properties required for organic fibre composites. While several authors have applied technology developed for E glass fibres to high-modulus Kevlar fibres it has generally been recognized that the organic fibre will require a different treatment.[9] There has been considerable research into the bonding mechanism of aramid fibres to epoxy matrices[9,10] and attention has recently been turned to extended-chain polyethylene. In particular, there have been developments of simple gas-phase treatments for the fibre to increase wettability and etch the surface.[11] There has also been an increased understanding of the surfaces of treated and untreated fibres through the use of sensitive analytical techniques such as X-ray photoelectron spectroscopy.[12,13] In this chapter the present status of research into the modification and control of the surface chemistry of extended-chain polyethylene fibres in order to improve the properties of the composite materials with thermosetting matrices will be examined.

2 HIGH-MODULUS POLYETHYLENE FIBRES—PREPARATION AND PROPERTIES

Unlike Kevlar fibres which are formed from intrinsically stiff rigid-rod aramid molecules pre-oriented in a liquid crystal solution, ultra-high-modulus polyethylene (UHMPE) fibres are obtained by drawing the polymer to produce chain orientation.[7] This requires exceedingly high draw ratios (in excess of 50) and the intrinsic entanglements of high molar mass polymers control the conditions under which this can be achieved.

Melt spinning and drawing is only possible for low molecular weight high-density polyethylene (HDPE) ($M_w \sim 10^5$) because of these entanglements, and this restricts the ultimate modulus to around 50 GPa. Another consequence of the low molecular weight of the melt spun material is the poor creep performance, particularly at elevated temperature.[11]

Entanglement concentrations in high molecular weight polyethylene have been reduced by dissolving the polymer ($M_w > 10^6$) at around 5–10 % wt in a solvent such as decalin or paraffin oil at 150°C. On cooling to below the melting point of HDPE a porous crystallized 'gel' is formed in which the disentangled state is maintained. This may now be spun and ultra-drawn at 120°C to achieve tensile strengths of ~ 3 GPa and moduli around 170 GPa. Gel spinning is the basis of the commercial ultra-high-modulus fibres Dyneema and Spectra.[8] These show superior creep properties to the melt spun material but at temperatures in excess of 120°C tensile creep is still a limiting factor.

In high-temperature curing epoxy resin composites (such as diaminodiphenyl sulphone cured systems), polyethylene cannot be used because the cure temperature (135–177°C) exceeds the temperature at which fibre shrinkage commences. There are, however, many amino cured epoxy resins as well as unsaturated polyester and vinyl ester resins which are suitable thermoset matrices for UHMPE composites. In composite materials the surface structure of the fibre is of particular significance. Studies of the compressive failure of organic fibres frequently show a core/skin morphology, with shear buckling of the outer fibre layer.[14,15] In Kevlar, this difference in properties between the skin and the core is believed to be due to differences in chain orientation, with the skin being more disordered due to quenching during spinning.[9] The behaviour of the skin of organic fibres may also be seen by microscopic examination of a single fibre bent into a loop. Aramids show compressive buckling on the inner circumference and tensile splitting on the outer[14] while ultra-oriented polyethylene shows only compressive buckling on the inner surface.[8]

Compressive buckling has been shown potentially to occur in composite material fabrication. If an unrestrained fibre is embedded in epoxy resin the compressive stresses due to thermal mismatch produce buckling of the fibre. The stresses set up at the interface may be seen through photoelasticity. Figure 1 shows an electron micrograph of UHMPE fibres (Spectra 1000) which have been embedded in a polyester resin. Compressive buckling has resulted in the formation of kink bands, possibly due to stresses set up at the interface by the shrinkage of the resin during cure. In this case the fibre had been plasma treated to increase interfacial adhesion, as discussed later (Section 8.2). Dobb *et al.*[14] have noted in aramids that TEM examination of kink bands has revealed microcracks and suggest that this would lead to a loss of tensile strength of the fibre.

Tensile failure of UHMPE fibres, in contrast to brittle carbon or glass, produces fibrillation with the crack in the failed fibre running for a long distance along the length of the fibre. This process, as for Kevlar, explains in part the excellent performance of the material at high rates of strain such as in ballistic

Figure 1. Scanning electron micrograph of Spectra 1000 ultra-high-modulus poly-ethylene fibre removed from an unsaturated polyester resin, showing kink bands due to compressive buckling of the fibre

impact because of the high fracture energy associated with the fibrillation process in a high-modulus material. However, all ultra-oriented fibres show a transverse tensile strength which is only a fraction of the longitudinal strength.[16] In UHMPE the transverse strength will be governed only by the weak van der Waals forces between the oriented fibrils in the crystalline region and by the residual entanglements in the unoriented zones between crystallites. These, along with chain ends, are the defects which limit the ultimate strength of the fibre in the draw direction, but are important in the surface reactivity of the fibre as they represent accessible polymer for carrying out chemical reactions, as will be discussed later.

3 INTERFACIAL PROPERTIES OF COMPOSITE MATERIALS

The analysis of the mechanical properties of unidirectional fibre-reinforced composites has traditionally been performed for high-modulus brittle fibres in a brittle matrix of lower modulus.[2] The ultimate strength of the composite is governed by the cumulative damage to the load-bearing elements. Since fibre strength in a real system will be variable, then it would be expected that there will be progressive failure of the weakest fibres, assuming all are equally loaded. This load, previously taken by the broken fibres, must be transferred through the matrix by shear stress to the remaining fibres. These remaining intact fibres must be well bonded to the matrix in order to restore mechanical equilibrium.

The fractured fibre is considered to have an ineffective load-bearing length and the ultimate strength is governed by the statistical accumulation of these isolated failed zones. Cumulation of these zones will lead to crack formation and catastrophic failure of the composite. If cracks could be prevented from propagating transverse to the fibre, the cumulation of damage in localized zones could be suppressed. This requires a *lowering* of interfacial bond strength between the fibre and resin, in contrast to the high bond strength required for maximum stiffness. This is just the situation for achieving fracture toughness of a composite.

In order to achieve high fracture toughness in a composite of two intrinsically brittle materials (e.g. glass and epoxy resin) it is necessary to exploit the large interfacial area for energy absorption and crack deflection by controlled debonding. This is traditionally demonstrated by a diagram such as Figure 2 which shows the various contributions to the total work of fracture of the glass–epoxy composite.[17] These contributions are:

(a) the fracture energy of the fibre and matrix,
(b) the work of debonding the fibre from the matrix,
(c) the work of pulling the fibre out of the matrix.

The last term is generally the dominant one in the work of fracture and the length over which pullout occurs is inversely proportional to the shear strength of the bond between the fibre and the matrix.

If the shear strength is so high that debonding could not occur, the crack

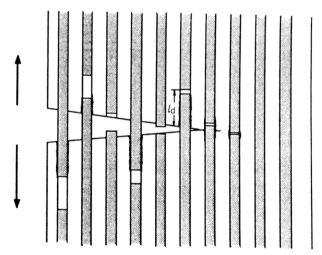

Figure 2. Debonding and pullout of fibres during crack opening in tensile deformation of a composite of a brittle fibre (e.g. glass) in a brittle matrix (e.g. epoxy resin). (Reproduced from Stachurski[17] by permission of Royal Australian Chemical Institute)

would propagate without any deflection, and the work to fracture would thus be the fracture energy of the two brittle components, which is negligible. In glass fibres, the use of coupling agents on the fibre is believed to achieve a compromise between good, stable bonding at low rates of loading in order to maintain strength and stiffness and debonding and pullout at high rates of strain to achieve toughness. This has been achieved more by a process of elimination than by design and the understanding of the micromechanics of this system came along after the technology for glass fibres was developed.[18]

When considering a composite in which the brittle fibre such as glass or carbon is replaced by the tough fibrillar aramid or oriented polyethylene, the relative importance of contributions to the fracture toughness of the composite, as discussed above, are changed. The fracture energy of the organic fibre is many times that of the brittle fibre and this will now make a major contribution to the total energy absorption. The fibrillation process in which longitudinal cracks open within the fibre offers a major mechanism for energy absorption at failure. The viscoelastic nature of organic fibres also offers energy absorption at strains well below failure. For example, hybrid composites fabricated with aramid or extended-chain polyethylene in the structure show superior vibration damping compared to pure glass fibre composites.[8] The intrinsic toughness of the fibre suggests that the interfacial properties of the organic fibre composite should be tailored to achieve maximum load transfer into the fibre. Thus debonding and pullout should not be the major energy-absorbing processes in extended-chain polyethylene and aramid composites, as they are in carbon or glass composites. Therefore, in tailoring the interfacial properties of these composites, a *maximum* bond strength should be achieved. Unlike the brittle fibre composites, this will not lead to damage localization and catastrophic failure under tensile loading since the fibrillation accompanying fibre failure leads to crack deflection. The crack must then reinitiate and this will be well away from the original advancing crack front.

4 EXPERIMENTAL METHODS FOR STUDYING FIBRE–RESIN INTERFACIAL PROPERTIES

The importance of interfacial properties in the performance of composite materials has resulted in a range of macroscopic, microscopic and molecular analytical methods being employed.[3] In any one system it is important that several techniques be used in order to determine the interfacial behaviour governing the engineering properties.

4.1 Macroscopic Tests

The most common mechanical test performed on unidirectional fibre-reinforced composite materials to assess interfacial properties has been the short-beam shear test[3] according to ASTM D2344. Interlaminar shear is achieved by performing a three-point bend on a sample with a small span-to-depth ratio. Frequently samples show compressive damage and it is not a valid test unless visual evidence of shear failure in the mid-ply region is obtained. A variation of this test is to use a four-point bend with a larger span-to-depth ratio.[19] The test cannot yield an absolute measure of bond strength but is particularly valuable for comparison between different surface treatments.

Transverse tensile strength tests should give a direct measure of interfacial adhesion in a unidirectional composite, but they are susceptible to any voids or processing defects in the sample.[3,16] A more controlled test is to determine the interlaminar fracture energy, G_{IC}, of a beam that contains a mid-ply starter crack. A transverse load is applied at the end containing the crack and the load required to extend the crack a measured distance (generally on the order of millimetres) is measured and used to calculate G_{IC}.[20] Bascom[5] has noted that when the fracture surface is examined in the electron microscope the actual failure mode can be masked because a fibre at the crack front experiences tensile stresses at the top of the fibre, changing to compression at the mid-plane. The fracture topography changes from apparent interfacial failure at the top to shear yielding of the matrix where the stress changes from tensile to compressive. Interpretation of the mode of interfacial failure from micrographs from these tests is thus difficult.

4.2   Microscopic tests

The difficulty in achieving a simple stress field in a macroscopic composite test has resulted in simplification of the test geometry to involve a single fibre embedded in the matrix. The first of these tests involved a single fibre totally embedded in a tensile bar of the cross-linked resin which has a failure strain greater than that of the fibre. On elongation, the local stresses will cause the fibre to fracture into segments of length between the critical length, l_c, and $1/2l_c$; the distribution of segment lengths depends on the strength distribution of the fibre. A statistical analysis allows the critical length l_c to be determined and a sheer stress, τ, at the fibre–resin interface to be estimated from

$$\tau = \frac{\sigma_u d}{2l_c} \qquad (1)$$

where σ_u = ultimate fibre strength and d = diameter. The stress distribution at the interface may be determined by examining the epoxy in polarized transmitted light. The resultant stress birefringence has been used to examine the locus of interfacial failure for surface modified carbon fibres in epoxy resins.[5]

Microscopic Raman spectroscopy of the embedded fibre has recently been used to measure the actual strain in the fibre through the shift in particular vibrational modes.[21] This may be measured with a resolution of a few micrometres along the length of an embedded carbon fibre and used to map the variation in τ with distance from the end of each of the fibre fragments during loading of the sample.[22] This method has been shown[23,24] to be applicable to a wide range of organic high-modulus fibres, including UHMPE, to assess interfacial adhesion with epoxy resins.

4.3 Pullout Tests

A more direct measure of the interfacial shear strength of an embedded fibre is the single-fibre pullout test, as shown in Figure 3(a). The fibre is either totally or partially embedded in a plug of resin to a depth, l, and then the maximum force, F_m, necessary to debond the fibre is obtained.[3] The frictional force on the fibre as it pulls through the socket is also obtained and gives an indication of the shrinkage forces during cure:[25]

$$\tau = \frac{F_m}{\pi dl} \tag{2}$$

Experimental difficulties arise at high interfacial bond strengths since the maximum embedded length, l_m, is governed by the ultimate tensile strength of the fibre:

$$l_m = \frac{\sigma_u d}{4\tau}$$

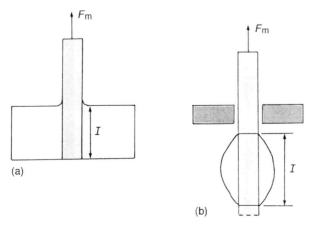

Figure 3. Arrangement for measurement of interfacial shear strength (a) by single fibre pullout at force F_m of an embedded length, l, in a resin plug and (b) by pulloff of a microdroplet of wetted length, l, through a knife edge

For a high bond strength (e.g. carbon–epoxy) this can result in embedded lengths which are less than 0.5 mm. Because of end effects where the fibre emerges from the resin plug, it is preferable to measure debonding force for a range of embedded lengths up to l_m and the slope of the plot of F_m versus l is used to obtain τ.[26]

An alternative, but challenging, method that has become popular recently is the microdroplet pulloff test.[27] This involves curing different size droplets of resin on the fibre and then gripping one end and measuring the force to pull the fibre away from the droplet when it is passed through a knife edge of spacing just greater than the fibre diameter (Figure 3(b)). The test need to be performed under microscopic observation to ensure a clean failure, but it overcomes the end effects of the fibre pullout test. The same analysis method is used to obtain τ.

One objection to all of the single-fibre test methods for interfacial adhesion is that they are far removed from an actual composite material and, in particular, effects that occur in fibre roving bundles such as incomplete wetout, fibre contact and void formation cannot be considered. A fibre bundle pullout test, identical to a single-fibre plug pullout test, provides an alternative.[27] However, if the incomplete wetout phenomena described above now occur, it is no longer possible to determine an interfacial shear strength, τ. The most accurate fibre bundle test method involves measuring pullout force as a function of embedded length and determining the slope of F_m versus l.[12] Capillary rise of the resin during wetout and cure will introduce an uncertainty in l that is significant at low values. This test forms a link between the microscopic and macroscopic test methods.

5 SURFACE CHEMISTRY OF ORGANIC REINFORCING FIBRES

Organic fibres obtained from the commercial fabrication process will have a range of surface finishing agents. The effect that these will have on the interfacial properties of the composite must be carefully assessed. The strongly hydrogen-bonded fibres such as Kevlar would be expected to have a high surface energy, based on the amide structural unit. As shown in Table 1, wettability studies on Kevlar aramid have given values from 43.7 to 47 mN m^{-1} which are greater than those for some carbon fibres.[9] The intrinsic surface energy of Kevlar may be masked by the finishing agents on the fibre which are not always removed by commercial scouring treatments.[28] Various surface treatments for aramid fibres have been researched in order to overcome these limitations and increase the bond strength with thermosetting matrices, particularly epoxy resins. Surface oxidation and etching treatments which have been shown to be successful with carbon fibres[6] led to damage of the aramid and a loss of tensile strength.[9] The introduction of amine groups on the surface by either ammonia plasma treatment[10] or a wet chemical reaction bromination followed by ammonolysis[29] increased the interlaminar shear strength of the epoxy composite without damaging the fibre. In contrast, the chemical reactions studied by Penn *et al.*[30] have resulted in no improvement in interfacial adhesion and it has been suggested[9] that this may be due to the weaker outer skin intrinsic to the skin/core morphology described earlier (Section 2).

Polyethylene surfaces have an extremely low surface energy, ~ 340 mN m (Table 1), so they will not be wetted by polar resins and will have no intrinsic reactivity with epoxies, unsaturated polyester or vinyl ester resins. The surface modification of polyethylene has been widely researched and the initial approach in the improvement in the bonding of extended-chain polyethylene fibres has been to apply techniques that are appropriate for unoriented polyethylene film. The development of surface treatments for organic reinforcing fibres as well as a better understanding of interfacial adhesion requires sensitive methods for the analysis of the functional groups on the surface of the fibre and their reactions with the matrix.

5.1 Surface Analysis of Fibres

Physicochemical methods of analysis are required to obtain a measure of the surface interactions between the reinforcing fibre and the matrix.

5.1.1 Contact Angles

The first requirement in achieving adhesion is that there be significant contact between the liquid resin and the fibre. The contact angle, Θ, that a liquid makes with a plane surface depends on the surface tensions for the solid and liquid, and a critical surface tension of wetting, γ_c, is often defined for a polymer such that a liquid with a surface tension lower than γ_c will spread on it.[31] The contact angle is a probe of the non-covalent interactions occurring between a liquid and the first monolayer of the polymer surface. These will provide contributions to the total work of adhesion and comprise dispersion or van der Waals forces, Lewis acid–base interactions (which include hydrogen bonding) and electrostatic interactions.

The direct measurement of contact angle for small-diameter fibres is difficult and two methods are commonly employed. In the first, the profile of a single drop on a fibre is measured using a goniometer and the contact angle is calculated.[32] This generally is considered accurate to 3° and is more accurate than the Wilhelmy method in which the force is measured as the fibre is pulled through the interface. The drop method has given surface energies for Kevlar fibres of 47 ± 1.5 mN m^{-1} and for an ultra-high-modulus polyethylene monofilament of 36 mN m^{-1}.[33] The differences between the two materials clearly reflect the almost total contribution to the surface energy of polyethylene of dispersion forces while Kevlar is strongly hydrogen bonded.

5.1.2 Inverse Gas Chromatography

The relative contributions of acid–base and dispersion interactions to the total surface energy may be determined by measuring the elution behaviour of probe molecules with different functionalities when the fibre is used as the stationary phase in a gas chromatograph.[34] The use of the technique to examine the nature of the surface contamination on Kevlar fibres has been reported[35] and the effectiveness of various cleaning techniques in achieving the highest surface energy for the fibre has been assessed. Rigorous solvent extraction was required to achieve the strong acid–base interactions to be expected from the amide groups of the aramid. Further information on the nature of the surface of organic fibres may be obtained by combining inverse gas chromatography with surface-sensitive spectroscopy such as XPS.[28]

5.1.3 Surface-sensitive Spectroscopy

While surface energy measurements such as contact angle and inverse gas chromatography probe the first monolayer of the fibre they do not provide chemical structural information about the surface. The techniques that provide such information in the first few monolayers are X-ray photoelectron spectroscopy (XPS or ESCA) and static secondary ion mass spectrometry (static SIMS).[35] In SIMS a focused ion beam (e.g. Ar^+, Xe^+) is used to sputter ions and neutral fragments from the surface at either a high rate so the surface is etched away (dynamic SIMS) or at a very low rate so that less than a monolayer is removed throughout the experiment (static SIMS). The latter experiment requires ion currents of ~ 1 nA cm^{-2} and is of more value in studying the surface of polymers because of their susceptibility to damage under dynamic SIMS conditions. The positive or negative ions are analysed by a quadrupole or time-of-flight mass spectrometer and the surface composition deduced from the parent ion masses and fragmentation patterns.[36]

Static SIMS has been used to study the surface of E glass fibres reacted with a methacryl functionalized silane to determine the optimum silanization conditions to achieve covalent bonding with the matrix.[37] While dynamic SIMS has been used to study the changes in silane structure with depth, such studies are generally unsuitable for the organic fibres because of polymer degradation. The advent of imaging static SIMS technology has opened up the possibility of mapping the failure surface of the fibre after interlaminar shear in order to determine the bonded regions.[35] One of the fundamental problems with SIMS is the difficulty in quantification, particularly compared to XPS, and for polymers in particular the following further problems have been noted:[36]

(a) the time dependence of certain peak intensities due to ion beam damage;
(b) changes in surface potential as charge builds up causing peak shifts and relative ion intensity changes.

The XPS experiment involves the measurement of the energy and intensity of photoelectrons ejected from a surface following X-ray photoionization. As applied to a high-modulus organic fibre, this allows the identification of all elements present other than hydrogen and helium, their oxidation state and the relative atomic concentration in the analytical volume.[38] Inelastic scattering of the photoelectrons restricts analysis to around three times the mean free path to the surface. While this depends on the kinetic energy of the photoelectron and thus on the X-ray energy of the source, it is typically 30–50 Å for non-metals such as the high-modulus fibres. This limited escape depth allows the surface composition as a function of distance from the surface to be measured. For flat surfaces, this is simply achieved by measuring the XPS signal at different emergent angles, with low (grazing) angles being the most surface sensitive.[35,38] However, with the highly curved surface of the small-diameter fibres this is not possible, since at grazing angles there will still be contributions from photoelectrons emitted normal to the fibre surface. An alternative approach uses the empirical relationship between the photoelectron escape depth and its kinetic energy.[38] An estimate of the depth distribution of an element may be obtained by either:

(a) using X-ray sources of high energy, so increasing the kinetic energy of the photoelectrons, e.g. a magnesium anode (1235.8 eV, K_α) gives a sampling depth of ~ 40 Å for the carbon 1s region while a titanium anode (4510 eV, K_α) would extend this to ~ 300 Å; or

(b) for a single anode energy, comparing the intensity of signals for two different lines of the same element having widely different binding energies and thus kinetic energies when photoionization occurs (e.g. oxygen 1s and 2s regions).

While elemental identification and quantification is relatively straightforward, the determination of the relative amounts of chemically similar functional groups is often difficult. In the study of organic fibre surfaces this is most frequently the carbon 1s region. X-ray photoelectron spectra[12] from the surface of an untreated and a corona discharge oxidized sample of commercial ultra-high-modulus polyethylene (Spectra) fibres in the carbon 1s region are shown in Figure 4. These spectra were obtained with a non-monochromatic Mg K_α anode which results in a linewidth of ~ 1.5 eV for the unoxidized carbon 1s band. The use of a monochromatic source produces a narrower instrumental line width (less than 1 eV) but causes problems with surface charging. The effect of surface charging may often be seen in the spectra of polymers since the binding energy is shifted by ~ 2 eV to higher energy of the carbon 1s band at 284.6 eV. The Bremmstrahlung from the non-monochromatic source assists in neutralizing the positive charge remaining on the surface of the insulating polymer following photoionization. This results in a stable

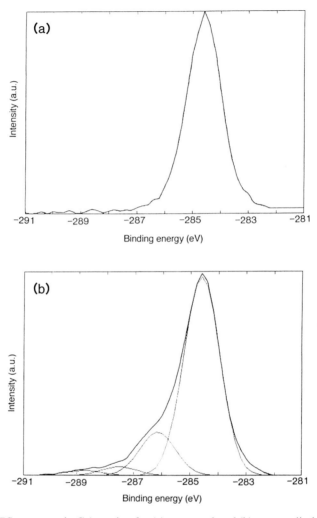

Figure 4. XPS spectrum in C 1s region for (a) untreated and (b) corona discharge treated Spectra UHMPE fibres. The peaks due to the oxidized carbon from the curve fitting of (b) are shown. (Reproduced from George and Willis[12] by permission of Carfax Publishing Company)

binding energy shift in the spectrum after a short time and no differential charging is observed.

It is seen from the spectrum of the UHMPE surface that the information of the type and concentration of the surface oxidation products is contained in the tail of the carbon 1s spectrum which lies to higher binding energy. It has been noted that the carbon 1s band shifts by ~ 1.5 eV for each increase in the oxidation state of carbon. Thus, based on expected binding energies of:

$$C-O: \quad 286.1 \text{ eV}$$
$$C=O \text{ or } O-C-O: \quad 287.6 \text{ eV}$$
$$O-C=O: \quad 289.2 \text{ eV}$$

curve fitting of the band may be performed. It is necessary to maintain a close tolerance in peak position and allow a band width of ~ 1.7 eV due to peak broadening on oxidation in order to obtain a meaningful curve-fitted spectrum as in Figure 4. By using the areas of each of the fitted peaks and sensitivity factors from standard compounds, the surface atomic concentrations may be obtained.

A particular problem that occurs in the XPS analysis of organic fibres but does not occur in the study of glass or ceramic fibres is the radiation damage to the surface from the X-ray beam or Bremmstrahlung. The dose rates may reach over 10 kGy min^{-1}, which is sufficient to produce radiation chemistry of the polymer in the time taken to acquire a multiplex spectrum (~ 20 min). This has been reported to occur in halogenated surfaces[39,40] and in oxidized UHMPE a loss in carboxyl groups was reported due to their decomposition to carbon dioxide under the X-ray source.[13] This was seen as a dramatic decrease in the oxygen content of the surface with analysis time, as shown in Figure 5.

Some of the problems of overlapping bands and radiation sensitivity of certain functional groups may be overcome by derivatization of the surface. The principle followed in XPS derivatization is to use a functional-group

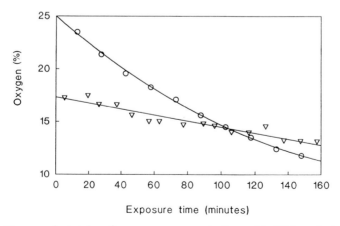

Figure 5. Decrease in total surface oxygen concentration with XPS analysis time due to polymer radiation chemistry for corona treated (∇) and oxygen plasma treated ($\bigcirc$) UHMPE (Spectra 1000) fibres. (Reproduced from Chappell *et al.*[13] by permission of Commonwealth of Australia)

specific reaction which results in another element being attached quantitatively to the surface.[35,40,41] By using known sensitivity factors for the new element, the surface concentration of the original functional group may be determined. If the new element has a higher photoionization cross-section there will be an increase in the overall sensitivity of the analysis. Table 2 summarizes some of the functional groups relevant to the surface analysis of high-modulus fibres which have been successfully analysed in this way. A similar approach has been used in the analysis of the surface of polymers by FT-IR reflectance spectroscopy, and it has been reported that it is only possible to obtain reproducible derivatization with gas-phase reagents.[42] Many of the reagents proposed involved solution reactions and it has been shown that surface restructuring may occur when prolonged solvent contact occurs. For reasons such as this, the surface of the polymer has been considered to be that which is accessible to reagents that are insoluble in the polymer when they are dissolved in a

Table 2. Some derivatization reagents for the XPS analysis of functional groups on the surface of high-modulus fibres[35,40,41]

Functional group	Reagent	XPS line	Analysed
$-\overset{\textstyle\|}{\underset{\textstyle\|}{C}}-OH$ Alcohol	$(CF_3CO)_2O$	F 1s	686 eV
	$(i\text{-Pro})_2Ti(acac)_2$	Ti $2p_{1/2}$	460 eV
$-\overset{\textstyle\|}{\underset{\textstyle\|}{C}}-OOH$ Hydroperoxide	SO_2	S 2p	164 eV
$-\overset{O}{\overset{/\backslash}{\underset{\|}{C}-\underset{\|}{C}}}-$ Epoxide	$(CF_3CO)_2O$	F 1s	686 eV
$-\underset{\overset{\|}{OH}}{C}{=}O$ Carboxyl	NaOH	Na 1s	1072 eV
	Tl O C_2H_5	Tl $4f_{5/2}$	122 eV
$-\underset{\|}{C}{=}O$ Carbonyl	$C_6F_5NHNH_2$	F 1s	686 eV
$-NH_2$	HCl	Cl 2p	200 eV
	C_6F_5CHO	F 1s	686 eV

non-swelling solvent. Even some problems have been reported with gas-phase reagents such as trifluoroacetic anhydride (TFAA) which will label epoxy and hydroxyl groups. Considerably higher levels of fluorine were observed than expected from the number of hydroxyl end groups present in a sample of polyethylene terephthalate.[40] This was attributed to reaction of the reagent with water in the polymer and the immobility of the resultant trifluoroacetic acid. This highlighted the problem of the reaction of derivatization reagents with additives and other chemicals present in the polymer and the interpretation of the subsequent XPS spectrum. Another problem noted with this reagent was the loss of fluorine with time of XPS analysis of the derivatized surface.[40] This may result, in part, from a dehydrohalogenation reaction under the X-ray beam. The limitations of XPS analysis of UHMPE due to the radiation chemistry of the sample were discussed earlier in this section. The reviews by Babitch[41] and Gillberg[40] provide a good summary of the advantages and limitations of surface derivatization in XPS analysis.

5.1.4   Luminescence Methods

In many surface treatments of fibres, the reactive environment such as a corona discharge or a gas plasma or the irradiation with UV or higher energy sources (electron beam or γ rays) results in the trapping of species in the structure which may subsequently react with the matrix during the fabrication of a composite. Species may include alkyl and alkyl-peroxy radicals, anions and cations, trapped electrons as well as thermally labile chemical groups such as epoxides, peroxides and hydroperoxides.[44] As discussed in Section 5.1.3, specific stable functional groups may often be analysed by XPS but the analysis of these transient surface species is often difficult. While electron spin resonance has the sensitivity to analyse bulk concentrations of species with unpaired spins of 10^{-8} mol dm^3, it lacks the necessary surface sensitivity.

Luminescence techniques have the necessary sensitivity to detect trace amounts of reactive species and one of the most common experiments is to heat the surface-modified polymer in an inert atmosphere or vacuum and observe the intensity of emitted light as a function of temperature.[45] Photon-counting photomultipliers and pulse-counting techniques have extended the sensitivity so that emission events of only a few photons per second may be quantitatively studied. If the fibre surface has been treated in an inert atmosphere there should be no labile oxygenated species and the luminescence is considered to arise by recombination of a trapped electron with a positive hole which is generally the cation from a photoionization event.[45] Spectral analysis has shown that the actual luminescent centre is often an impurity such as an aromatic or carbonyl group.[46] In a study of the thermally generated luminescence following electron beam irradiation of the surface of low-density polyethylene, it was found that there was a thirtyfold increase in emission intensity when the polymer was irradiated in air compared to nitrogen, and the maximum in the glow curve occurred at a much lower temperature.[47] The emission from the air-irradiated sample was attributed to the thermal decomposition of peroxides and hydroperoxides formed in a free-radical chain reaction initiated by the electron beam irradiation. This and other studies have shown a qualitative relationship between integrated luminescence intensity and hydroperoxide concentration as determined by calorimetric or titrimetric analyses.[47,48] The mechanism for this chemiluminescence from the thermolysis of hydroperoxides has been considered to be either:

(a) unimolecular decomposition resulting in a cage rearrangement of the alkoxy radical and hydroxyl radical to produce an electronically excited carbonyl group and water; or
(b) a bimolecular reaction between alkyl peroxy radicals produced by abstraction of the hydrogen atom from hydroperoxides by the mobile hydroxyl radical.

This termination reaction of peroxy radicals proceeds through a six-membered intermediate and produces an alcohol, singlet oxygen and an excited carbonyl group in its first triplet state.[48]

From studies of the effect of radical scavengers on the emission, a cage mechanism appears unlikely so that mechanism (b) is favoured.[48,49] In a study of the very early stages of UV irradiation of polypropylene it has been shown that the luminescence method is able to detect hydroperoxide formation and

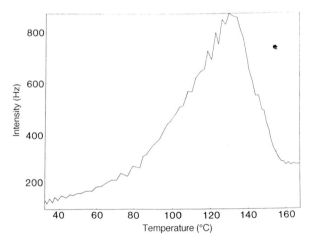

Figure 6. Chemiluminescence emitted from hydroperoxidized Spectra UHMPE yarn when heated in an inert atmosphere. A repeat temperature scan resulted in no further luminescence emission. (Reproduced from George and Willis[12] by permission of Carfax Publishing Company)

photodecomposition which cannot be observed by XPS.[49] The only obvious limitation to the technique is the lack of surface specificity in optically transparent polymers. Figure 6 shows the intensity of chemiluminescence as a function of temperature when an UHMPE fibre containing hydroperoxide groups is heated in nitrogen.[12] The area under the curve is proportional to the hydroperoxide concentration.

6 SURFACE CHARACTERISTICS AND MODIFICATION OF UNORIENTED POLYETHYLENE

The low surface energy of polyethylene has long been recognized as a problem in achieving adhesion to films, fibres and moulded articles. Brewis and Briggs[31] have presented a summary of the successful surface treatments for polyethylene to enable operations such as printing of film to occur. These involve either a gas- or solution-phase oxidation of the surface and XPS has shown that the minimum treatment necessary involves insertion of around 4 oxygen atoms per 100 carbon atoms over the XPS sampling depth of ~ 50 Å. The treatments commonly employed are corona discharge, flame treatment, chromic acid etching, inert gas bombardment, plasma treatment and photo-oxidation.

One question which many investigations have addressed is the relative importance of the physical and chemical modifications to the surface that these treatments produce.[31,35,51–55] These include:

(a) an increase in surface polarity which enhances wetting of the surface and introduces specific interactions such as hydrogen bonding,
(b) removal of a weak boundary layer and other surface contaminants,
(c) roughening, enabling complete spreading and mechanical keying,
(d) cross-linking of the first 100 nm,
(e) chemical bonding to new functional groups inserted into the polymer surface.

From a consideration of the surface energetics of polyethylene, the magnitude of the dispersion component of the surface energy alone is sufficient to achieve adhesion to the substrate by an epoxy resin or similar adhesive.[31] The major problem appears to be the failure of the adhesive to wet the surface and consequently all of the successful treatments result in a lowering of the contact angle. This is generally accompanied by oxidation of the surface, so the introduction of polar groups leads to an increase in surface energy through the electrostatic and acid–base interactions which may now occur. It is thus impossible to separate out which is the most important component in achieving adhesion, although modern surface spectroscopy has enabled direct analysis of the result of a surface treatment to be studied.[35] For example, it was reported that the treatments which eliminated a weak boundary layer (ion bombardment and UV radiation) did not produce oxidation as determined by ATR-IR.[56]

However, studies using XPS have shown that oxidation of the outer 50 Å has occurred which could not be detected by ATR-IR.[31] In gas-phase treatments such as ion bombardment, long-lived radical species in an inert atmosphere will be readily scavenged by oxygen on exposure to air, resulting in the formation of peroxidic and carbonyl species. Similarly, reports of achieving a highly crystalline polyethylene surface which produced a high bond strength with an epoxy resin by first melting the polymer against aluminium[57] have been shown by XPS to be due to oxidation[58] rather than any property of the *trans*-crystalline surface.

Another important point in surface studies of a polymer such as polyethylene, which is above its glass transition temperature (T_g) at room temperature, is the rearrangement of the surface functionality with time. Since there will be a driving force to minimize the surface energy, the polar groups produced at the surface may be progressively returned to the bulk. This has been observed by angular resolved XPS and contact angle measurements and the rate will depend on the conformational freedom of the polymer chain and thus T_g.[40] It has been shown, however, that when the surface is placed in contact with a high surface energy liquid, the polar groups may reorient to the surface.[40] The important factor is thus the time taken for this reorientation in relation to the cure time of the adhesive. This may be overcome in the first place by ensuring that a minimum time elapses between surface treatment and bonding, and in most commercial treatment of polyethylene the oxidation step immediately precedes printing, bonding, etc.

Recent surface spectroscopic studies of polyethylene following surface treatment by acid etching, corona discharge, direct plasma and remote plasma have provided a greater understanding of the stability of the surface, the concentration and distribution of functional groups and their reactivity. Erikkson *et al.*[55] have examined the effect of strong chemical oxidants on low-density polyethylene using XPS, including derivatization, to separate the type of functional group produced by $KMnO_4/H_2SO_4$, $KClO_3/H_2SO_4$ and $K_2Cr_2O_7/H_2SO_4$. The first was the most powerful oxidant (as expected from its standard reduction potential) and the groups observed were: hydroperoxide, hydroxyl, carbonyl, ester, sulphonate and carboxyl. By annealing the surface at 80°C, migration between the surface and the bulk was accelerated and a smoother and chemically homogeneous surface resulted. By controlling the ionic state of the surface by pH or water contact at annealing it was possible to achieve a surface with a higher sulphonate group concentration.

The process of corona discharge treatment of polyethylene in air has been widely studied and Gerenser *et al.*[53] have compared XPS and contact angle data to observe the effect of surface ageing for periods up to 75 days. The major change occurred between 14 and 28 days and represented a reorientation into the bulk of oxygen species from the outer 5 Å. These may be low molecular weight species, since water washing after discharge treatment produced a surface that was stable with time. This was more pronounced at higher powers and water washing left ∼ 10% incorporated oxygen present as hydroperoxy, hydroxyl, carbonyl, epoxy and carboxyl groups, as shown in Table 3. Some evidence was presented for the hydroperoxy group being located away from the surface in an environment that would prevent it from reacting. However, it is interesting to note that Iwata *et al.*[59] were able to initiate the polymerization of acrylamide on the surface of a corona discharge treated high-density polyethylene film when the hydroperoxide was decomposed by the redox catalyst Fe^{II}. Some surface restructuring was apparent with storage times since the graft yield decreased to only 20% of the initial value, after 12 days storage. Chemical analysis of peroxide showed that the total concentration had, however, decreased only to 60% of the initial value, so there was clearly a significant concentration of peroxide species that were unavailable for grafting.

Gerenser[54] has presented a comprehensive study using angular resolved XPS of the species produced on treatment of high-density and low-density polyethylene by argon, nitrogen and oxygen plasmas. Plasma-modified surface were stable for greater than 72 hours in the vacuum chamber but on exposure to the atmosphere immediate changes were observed. The argon plasma treated surface which had shown no change in structure immediately adsorbed 2% oxygen. The nitrogen plasma treated surface lost 40% nitrogen, probably due to a hydrolysis reaction of an imine to an aldehyde and evolution of ammonia. Part of the change in surface composition may have been due to the formation of a hydrocarbon contaminant overlayer. This was seen for the case of the oxygen plasma treated sample which, because of its high surface energy, was contaminated within 30 seconds exposure to the atmosphere, producing an apparent decrease of 25% in the oxygen concentration by XPS. When these effects are eliminated, the extent of surface modification may be assessed. It was found that all three plasma treatments produced cross-linking to a depth of 1–3 μm, which was most probably due in part to the vacuum UV radiation

Table 3. Surface functional groups on corona discharge treated polyethylene containing ∼ 10% oxygen

Functional group	Concentration (%)	
	Initial	Water washed
Hydroperoxide	1.1	1.3
Alcohol	1.8	1.2
Carbonyl	1.1	0.8
Epoxide	0.9	0.4
Carboxyl	1.6	0.8
Total [O]	9.2	6.6

produced in the plasma. The maximum amount of oxygen that could be introduced was 20%, and this was located within 10–25 Å of the surface. The predominant species was ether (or ethoxide) followed by carbonyl and ester (or carboxyl) groups.

In order to eliminate some of the factors present in a direct plasma experiment, Fouerch and McIntyre[60] have studied 'remote' plasma treatment in which only species with lifetimes greater than 1 μs were able to reach the polyethylene surface and react. In addition to eliminating sputtering and vacuum UV irradiation effects, the reaction rate was reported to be much higher than for the direct plasma treatment. For example, a surface atom concentration of oxygen of 18 $\pm$ 1% was achieved in 1 s compared to 30 s for the direct plasma under comparable conditions. This may reflect the cross-linking occurring in the direct plasma which will compete with the available sites for oxidation. Angular dependent XPS showed hydroxyl and ether groups were at a greater depth than carbonyl and carboxyl (or ester) groups, which was not observed from the direct plasma. Comparison of the plasma technique with corona discharge and ozone oxidation showed an oxygen content from 15 to 18%, with similar functional groups being produced. These values are well in excess of the minimum 4% needed to achieve an increase in bonding to a polyethylene surface.[31]

In many applications it is the durability of the adhesive bond when exposed to environment extremes that is important. Many high surface energy environmental agents such as water may disrupt the intermolecular forces and lower the bond strength. For this reason coupling agents are frequently used which chemically bond to the surface and to the adhesive, reproducing a durable bond.[18] Functionalized silanes have been widely employed for controlling adhesion to substrates containing hydroxyl groups such as glass, but clearly polyethylene must be surface treated before this route may be employed. The same effect may be achieved directly by choosing a surface treatment that will produce functional groups which will bond to the adhesive, i.e. the treatment must be specific to the known chemistry of the adhesive to participate in the cross-linking reaction or polymerization process. This approach is similar to that which was demonstrated by Iwata *et al.* in which the polymerization of acrylamide was initiated by hydroperoxide groups from a corona discharge.[59] The resultant polymer will be grafted to the surface with an ether linkage as a permanent chemical bond.

As will be discussed later, this approach is being pursued in optimizing the interfacial adhesion of ultra-high-modulus polyethylene fibres. The challenge is to achieve sufficient modification of the surface of the fibre without affecting the mechanical properties. While in principle it should be possible immediately to translate the surface treatments on polyethylene film to the unoriented fibre, it has been found that the surface chemistry and reactivity of the fibre is affected by the extent of orientation of the polymer.

7 THE EFFECT OF ORIENTATION ON THE OXIDATION OF POLYETHYLENE

Studies of the kinetics of oxidation of polyolefins as a function of the degree of orientation have shown a decrease in the rate constants for propagation and termination and an increase in the induction period.[61,62] This was ascribed to a decrease in the conformational mobility of the polymer chain as orientation increased. In the case of polyethylene at a draw ratio of 12 the kinetic chain length was reduced from 6 to 4 for γ-initiated oxidation at 22°C.[62] The most dramatic effect was on the oxidation product distribution. There was a fivefold decrease in the yield of hydroperoxide and a twofold increase in the yield of carbonyl groups. This suggested that an induced decomposition of hydroperoxide was occurring in a chain process involving the macro alkyl radical. The lower diffusion coefficient and solubility of oxygen in the highly crystalline oriented polymer may have allowed induced decomposition to compete with radical scavenging by oxygen. When the thermally initiated oxidation was measured at 130°C the kinetics of the drawn and undrawn polymer were indistinguishable, reflecting the increased conformational mobility at the elevated temperature leading to relaxation of the oriented chains. There were still differences between the properties of the oriented polymer and the isotropic material, with a higher molecular weight and greater strength retained after oxidation, indicating that a small fraction of the polymer chains remains in an extended conformation, so resisting oxidation.

The above observations relate to bulk properties and it might be expected that there would be a much lesser effect on the surface. However, in a study of chromic acid oxidation of the surface of oriented polyethylene[33] it was found that there was an increase in the chemical stability of the surface with an increase in the draw ratio. Of particular interest was the observation that the surface energy of polyethylene increased to a value of 46 mN m^{-1} with a draw ratio of 14.3. This means that the oriented fibre would be wetted by an epoxy resin ($\gamma_L = 43$ mN m^{-1}) while an unoriented film would not. This was attributed to an increase in the dispersion component of the surface free energy, but no surface analysis was performed to determine whether oxidation had occurred during the drawing process, which was performed at 120°C in glycerol.

These results indicate that the surface and oxidation properties of drawn polyethylene are greatly different from unoriented polymer and it is not possible to extrapolate directly from the data for film to achieve optimum surface treatment.

8 STRATEGIES FOR ACHIEVING OPTIMUM INTERFACIAL ADHESION BETWEEN UHMPE AND THERMOSETTING RESINS

The consideration of the micromechanics of brittle thermoset matrices reinforced with ultra-high-modulus polyethylene fibres (Section 3) leads to a requirement for a maximum bond strength between the fibres and the resin to achieve a balance of strength and toughness in the composite material. The techniques available for modifying the surface of polyethylene (Section 6) may be applied to the high-modulus fibres with due regard to the decreased reactivity of the oriented polymer. Since the application of the fibre as composite reinforcement requires maintenance of a high ultimate tensile strength, the surface treatment should not damage the fibre.

8.1 Melt-spun UHMPE

Ladizesky and Ward first recognized the problems of the low surface energy and chemical inertness of ultra-high-modulus polyethylene fibres in fabricating composites with epoxy resins and undertook a study of a model system consisting of melt spun monofilaments of >0.2 mm diameter.[63] These were subjected to draw ratios of 8, 15 and 30 at 120°C and the interfacial adhesion determined by the fibre pullout method (Section 4.3). Two surface oxidation methods were compared—chromic acid oxidation and a direct oxygen plasma—and the surfaces of the monofilaments were examined before and after pullout from the plug of epoxy resin, using scanning electron microscopy (SEM). In some cases the pullout groove in the resin was examined after sectioning, to determine the locus of failure.

The plasma treatment consisted of 10 min in a reaction chamber capacitively coupled to a 13.56 m Hz generator operating at 10 W. The reactive gas was oxygen flowing at 10 cm³/min. This treatment produced a cellular surface on the monofilament with pits of up to 4 μm diameter (Figure 7). The pullout force after treatment increased twelvefold, resulting in an apparent interfacial shear strength, τ_a, of 5 MPa (ignoring the change in surface area on pitting) for the most highly oriented monofilament. However, this level of improvement occurred with a 50% decrease in the ultimate tensile strength of the filament. A linear inverse relation was observed between the ultimate tensile strength and interfacial shear strength, τ_a, for both plasma and acid treatment (Figure 8).

Pullout of the plasma-treated monofilament resulted in shearing away of the cellular surface so that failure at the highest values of τ_a was within the polyethylene. The resin had totally wet out the fibre since, after dissolution of the polyethylene remaining in the resin groove, a replica of the surface in positive relief was obtained. On the basis of these observations it was concluded that mechanical keying was the principal factor in achieving a high bond strength. This was supported by the much lower values of τ_a of 1.4 MPa achieved by chromic acid treatment (60 s at room temperature) in which no surface pitting occurred. Failure on pullout occurred at the interface and there was only a decrease of 13% in the ultimate tensile strength. It was of interest that, at a high draw ratio, perfect wetout of both untreated and treated monofilaments occurred. Clearly, wetting alone is insufficient to achieve optimal

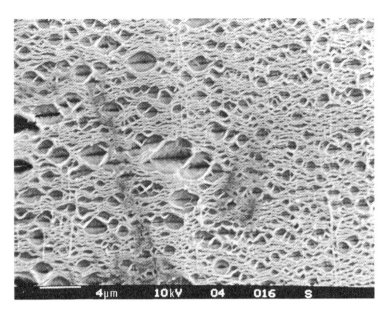

Figure 7. Scanning electron micrography of the surface of a monofilament of melt-spun UHMPE after oxygen plasma treatment, showing heavy surface pitting. (Reproduced from Ladizesky and Ward[63] by permission of Chapman and Hall)

interfacial adhesion in this system, which would be the case if dispersion forces had provided sufficient interaction, as suggested by Brewis and Briggs.[31] Unfortunately, surface analysis such as XPS was not reported for these samples to determine the level of oxidation of the polymer.

These model studies were extended by the same authors to the composite material, using a commercial prototype melt-spun multifilament yarn with a draw ratio of 30.[64] The plasma treatment conditions were based on those for the monofilament and were kept below the threshold for significant damage. Special techniques were developed to enable the continuous treatment of yarn as it was wound from a bobbin into the reactive zone of the oxygen plasma. The conditions employed for the plasma generator were similar to those for monofilament except that the power was 60 W. The effectiveness of the improvement in interfacial adhesion was measured by the interlaminar shear strength (ILSS) of sections of the composite (Section 4.1). This showed an increase in the ILSS ranging from 25 to 80% depending on the method of composite lay-up. Scanning electron microscopy of the fibres showed a high density of cracks which were of length $\sim 1 \mu m$ compared to the larger pits ($\sim 4 \mu m$) on the monofilament. This resulted in a loss in tensile strength of the composite of $<10\%$. Very good wetout of the fibres was observed and the locus of failure involved cohesive failure of the polyethylene.

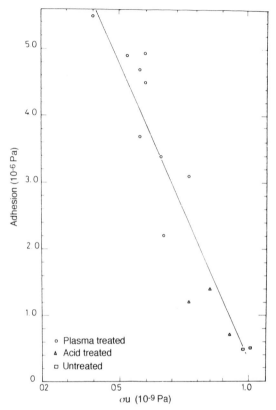

Figure 8. The interdependence of fibre-matrix adhesion and ultimate tensile strength (σ_u) of the fibre for various surface treatments of UHMPE monofilaments. (Reproduced from Ladizesky and Ward[63] by permission of Chapman and Hall)

Environmental exposure to water at the boil showed a minimal effect on the stability of the interface with a reduction of $\sim 19\%$ in ILSS after 3 days, indicating a retention of the strong interfacial adhesion. The only negative effect of the increased adhesion appeared on the impact strength of the composite. This showed a decrease for the first and successive impacts on plasma-treated fibres. Optical examination showed that the damage in these samples was confined to a much smaller area than in the untreated material. This suggested that debonding or matrix cracking made a small but significant contribution to the total fracture energy. However, the major component was the plastic deformation of the fibre.

This initial study of interfacial adhesion in epoxy resin–UHMPE composites was extended to polyester resins[65] and hybrid composites of UHMPE with glass and carbon fibres.[66] The ILSS of the polyester composite with plasma-treated fibre showed an increase of 35% compared to the untreated control. This may be interpreted as an indication that the increase in adhesion resulted from mechanical 'keying' effects rather than specific chemical interactions, since the cure chemistry of the epoxy and polyester resins are quite different and would require different surface functionalities for chemical bonding.

While these studies have demonstrated the general applicability of oxygen plasma treatment in increasing interfacial adhesion between UHMPE and epoxy or polyester resins, the actual mechanism was not considered in any detail. Nardin and Ward[33] reinvestigated the monofilament of melt-spun UHMPE with the aim of determining the relative importance of the contributions to the total interlaminar shear strength, τ, of: physicochemical (i.e. non-bonding) interactions, τ_{PCI}; mechanical keying, τ_M; and chemical bonding, τ_{CB}. In addition to chromic acid and plasma treatments in oxygen, the monofilament was annealed before treatment and was subjected to a water vapour plasma to insert hydroxyl groups, a functionalized silane treatment and sulphuric acid at $70°C$ or a borane reduction in order to change the concentration of hydroxyl groups that were proposed as sites for chemical bonding to the epoxy resin.

It was found that annealing at $110°C$ for 20 hours prior to plasma treatment in oxygen decreased the size of the surface pits while the sulphuric acid treatment increased their size and changed their shape. The effect of pitting on the total interfacial area was calculated and the remaining contribution related to the mechanical keying of the adhesive to the surface. This was shown to contribute a maximum 3 MPa to the total value of τ.

The physicochemical contribution, τ_{PCI}, was determined by measuring contact angles of epoxy resin and glycerol on the monofilament (Section 5.1.1) to determine the dispersion γ_s^D and polar γ_s^P contributions to the total surface energy of the treated polyethylene. As noted before, γ_s^D increases with the draw ratio so total wetout by the epoxy resin occurs for the ultra-high-modulus material. The value of τ_{PCI} is estimated as ~ 2 MPa, which is the value calculated from the pullout force of an annealed, plasma-treated monofilament which is effectively unpitted (i.e. $\tau_M = 0$). The number of molecular interactions, assumed to be hydrogen bonds, was estimated to be one per square nanometre.

Since the number of available carbon atoms on the surface of polyethylene is estimated[31] as 5×10^{14} cm^{-2}, i.e. 5 carbon atoms/nm^2, this means the surface concentration of oxygen is $\sim 17\%$ if the major oxygenated species is —OH as suggested by Nardin and Ward.[33] As discussed in Section 6, this is close to the maximum concentration of oxygen which was analysed on the surface of plasma-treated polyethylene films by XPS ($18 \pm 1\%$). This was achieved in a much shorter treatment time with no indication of surface pitting. As also noted in that section, the minimum surface oxygen concentration to achieve sufficient increase in surface energy for adhesive bonding is 4%.[31]

While Nardin and Ward do not consider that direct chemical bonding is a major component of the bond strength achieved between epoxy resin and oxygen plasma treated monofilament, they consider several strategies for increasing the chemical bonding contribution, τ_{CB}. The water vapour plasma as well as the borane treatment were expected to increase the surface hydroxyl group concentration, in the first instance by direct insertion of $\cdot$OH from the water plasma and in the second by BH_3/THF reduction of any carboxylic acid groups on the surface. These hydroxyl groups should then be available for condensation with a functionalized trimethoxy silane, which was then bound to the surface of the monofilament. The functional group, which in the system chosen was an epoxide, was then able to participate in the cross-linking reaction of the epoxy resin by the amine curing agent. This process achieved a value for τ_{CB} of 1.7 MPa.

It was noted that it was difficult to separate the contributions to τ_{CB} and τ_{PCI} by the experiments undertaken. Another factor that was not considered by the authors is the known catalytic effect of hydroxyl groups on the cure reaction of epoxy resins.[67] Depending on the cure conditions, the hydroxyl groups may also undergo etherification reactions with the epoxy resin to produce a direct chemical bond to the surface.[68] Both of these could affect the interfacial shear strength, τ, through increasing τ_{CB} rather than the assumed τ_{PCI}.

8.2   Commercial Gel-spun UHMPE Fibres

The earlier papers of Ward and coworkers discussed above have focused on the melt-spun UHMPE from low molecular weight polymer. However, the commercial fibres Dyneema and Spectra are gel spun from high molecular weight polymer (Section 2). The following studies of UHMPE surface modification and adhesion are concerned with the gel-spun, commercial fibres. Most attention has been paid to gas-phase plasma treatment, although the most recent papers are concerned with specific surface modification with a view to achieving chemical bonding with particular resin types.

Ward and coworkers extended their earlier studies of melt-spun monofilaments and fibres to include a wider range of low molecular weight commercial melt-spun fibres as well as a high molecular weight gel-spun fibre—Spectra 1000.[69] It was noted that the ILSS of Spectra composites was much lower than those from melt-spun material, but this was related to the differences in interfacial surface area due to the high diameter of the gel-spun material (36 μm). This was supported by a linear relation being observed between ILSS and the inverse fibre diameter. Oxygen plasma treatment produced an increase in ILSS for all fibres which reached a plateau after ~ 60 s. While this produced a decrease in the contact angle between epoxy resin and the Spectra fibres from 8–11 to 3–5°, the enhancement in adhesive bonding was not due solely to increased wettability from the surface oxidation. Rather it was attributed to the removal of the low molecular weight material believed to be rejected to the fibre surface during solidification, resulting in a weak boundary layer. This is similar to the proposed boundary layer on unoriented polyethylene (Section 6) which has, however, been questioned as being of significance in the ultimate achievable bond strength. Clear evidence was presented for cross-linking of the Spectra fibre surface by the oxygen plasma as a gel fraction was obtained that increased with time of treatment. The gel fraction increased for treatment times over 120 s whereas the ILSS showed no further improvement after 60 s. It is difficult to determine the significance of plasma-induced cross-linking in achieving an improved bond strength since the treatment also produced pitting of the fibre surface. The depth of the pits was much less than on the monofilaments (Figure 7) and probably reflected the greater oxidative stability of the higher modulus fibres. The pits were not considered to be as significant in achieving a high value of τ as in monofilaments, and may in fact be detrimental by lowering the fracture energy of the fibre.[69]

Kaplan *et al.* prepared unidirectional and woven fabric composites of Spectra 900 with an amine cured epoxy.[70] The yarn was treated in a continuous process by passing through a transition zone and into the RF (radio frequency) plasma reaction area and then through another transition zone and back to the atmosphere. The fabric was treated by a roll-to-roll process within the chamber which was operated at 250–800 mtorr and a power less than 350 W, in order to minimize fibre damage. No details of exposure time to the plasma or the atmosphere in the chamber were given but it is presumed to be oxidative.

The ILSS (Section 4.1) of the unidirectional composite increased from 8 to 31 MPa after fibre plasma treatment, in contrast to a value of 18 MPa for corona discharge treated fibres. The woven fabric composites showed more modest increases from 5–6 MPa, untreated, to 13–15 MPa, treated, depending on the type of weave. This was gained at the expense of a 10% decrease in the tensile strength of the fibre. The treatment also produced an increase in tensile modulus and a decrease in elongation at fail, suggesting that cross-linking over the first few hundred angstroms was occurring. This is in agreement with other authors[69] and is probably a consequence of the UV generated in the plasma. No data in support of a mechanism for interfacial adhesion enhancement was presented, and it was assumed that, on the basis of the ready wetout of the modified fibres and studies presented in the following paper,[71] that hydrogen bonding and mechanical interlocking, as proposed by Ward,[33,63] were occurring. In the second paper in this series[71] single-fibre pullout as well as ILSS measurements on unidirectional composites of plasma-treated Spectra 900 were presented, along with electron microscope and diffuse reflectance FT-IR studies of the surface before and after bonding. The treated fibres showed micropitting with enhancement around kink bands (formed by compressive buckling, see Section 2 and Figure 1) which appear to act as stress concentration points. The FT-IR study aimed to examine the differences in the surface composition before and after coating the treated fibre with epoxy resin. The plasma-treated fibre showed the presence of surface ketones, ethers, esters and secondary alcohols and it was claimed that there was a decrease in the latter functional group concentration during cure of the epoxy resin. No reaction mechanism between the alcohol and epoxy resin was proposed, but it is generally agreed that the only direct cross-linking chemical reaction would be etherification.[68] However, this reaction produces a secondary alcohol group for every one consumed, since the oxirane, on ring opening, gives an alcohol group, and there should be no net decrease in secondary alcohol concentration, as was observed. Similarly, if the reaction is proceeding through the usually accepted autocatalytic route, the secondary alcohol concentration as detected by FT-IR diffuse reflectance must increase.[67] These results must therefore be held in doubt and there is no clear evidence for direct chemical bonding of epoxy resin to plasma-treated UHMPE fibres. It is also noted that diffuse reflectance is not a surface-specific technique, as is XPS.

One of the few studies on UHMPE to use a plasma other than an oxidizing plasma is that of Holmes and Schwartz[44] in which an ammonia plasma was used to aminate the polyethylene surface. Ammonia plasmas have been successfully used for increasing the surface energy and adhesive bonding of polymethyl methacrylate to UHMPE short fibres[72] and Kevlar fibres to epoxy resins.[10] The UHMPE used in the study of Holmes and Schwartz was a plain woven fabric of Spectra 900 and the plasma treatment was carried out at 1 torr and powers from 50 to 150 W. Treatment time ranged from 1 to 10 minutes. Assay of the surface was performed by the contact angle of water droplets and by a dye analysis of the amine groups introduced into the polymer surface. The mechanical strength of the bond formed with an epoxy resin was determined by a T-peel test (ASTM 1876-72) but it was noted that the results were largely qualitative since the test was unsuited to the geometry of the fabric. The amine group concentration obtained was anomalously high and no XPS analysis was used to confirm the surface amination. The qualitative trends indicated that there was an increase in adhesive bonding as a result of amination and this was attributed to the enhanced wettability of the surface rather than chemical bonding, as suggested by Allred *et al.* from a similar study of surface-aminated Kevlar fibres.[10]

A more comprehensive study of the enhancement of adhesion of UHMPE to epoxy resins by both oxygen and ammonia plasma treatment has recently been reported.[13] The comparison of the two treatments allowed chemical bonding effects to be more readily recognized. Spectra 1000 woven fabric was treated in a capacitatively coupled commercial plasma reactor (Plasmaprep 500 XP) at 13.56 MHz and at 60 W (O_2) or 100 W (NH_3), 250–300 mtorr with a gas flow rate of 20 cm^3/min. The surface of the polymer was studied by XPS and a dye assay technique similar to that of Allred *et al.*[10] and Holmes and Schwartz[44] following NH_3 plasma treatment. The effect of treatment time on

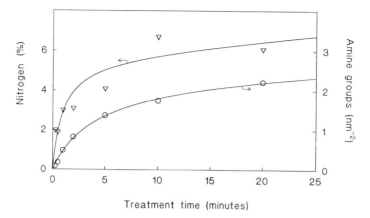

Figure 9. The increase in concentration of amine groups on the surface of Spectra UHMPE fibres subjected to an ammonia plasma. The amine concentration was determined by XPS (∇) and dye assay ($\bigcirc$). (Reproduced from Chappell *et al.*[13] by permission of Commonwealth of Australia)

surface composition is shown in Figure 9, and it is seen that there was an initial rapid increase in nitrogen content of the surface, followed by a slower increase with time. The concentration of amine groups by dye assay followed a similar curve. It is of interest to compare the magnitude of the two analytical results. For example, from Figure 9, after 15 minutes the surface contained 6% N by XPS and there were two NH_2 groups per square nanometre. If the number of available carbon atoms on the surface of polyethylene is 5 C atoms/nm^2, as indicated by Brewis and Briggs,[31] the amine group dye assay should correspond to an XPS analysis of 40% N rather than the 6% N observed. However, it has to be recognized that the XPS analytical volume corresponds to a depth of ~ 40 Å ($\sim 3\lambda$, the photoelectron escape depth), while the bulky azo dye may be restricted to just one or two monolayers. If amination occurs only in the outer layers of the ultra-high-modulus polyethylene, as indicated by the rapid plateau reached during plasma treatment (Figure 9), this will result in a low XPS analysis. This also assumes that all of the nitrogen is incorporated as amine and may be protonated and assayed by the azo dye. The reactive species in the ammonia are expected to be $\cdot NH$ and $\cdot NH_2$, so this would seem a reasonable assumption.

Primary and secondary amine groups will open the epoxide ring of a bisphenol A-based epoxy resin at low temperatures,[67] so the amine groups attached to the polyethylene surface should participate in the cure reaction during fabrication of the composite material:

$$RNH_2 + \underset{O}{CH-CH_2-CH_2-OPh} \longrightarrow \underset{\underset{OH}{|}}{RNHCH_2CHCH_2OPh}$$

Table 4. Interlaminar shear strength of DGEBA–DDM epoxy resin composites reinforced with Spectra 1000 UHMPE that had been subjected to corona discharge, oxygen plasma and ammonia plasma[13]

Treatment	ILSS(MPa)	Fibre content (wt %)
Untreated	5.7 ± 0.3	65.6
Corona	7.0 ± 0.8	64.8
2 min O_2 plasma	6.6 ± 0.6	64.7
1 min NH_3 plasma	11.1 ± 0.3	62.1
2 min NH_3 plasma	11.8 ± 0.6	62.4
10 min NH_3 plasma	11.8 ± 1.9	64.6

where R in this case represents the ultra-high-modulus polyethylene backbone. Chappell *et al.*[13] consider that this mechanism is operating, since ammonia plasma treated UHMPE fibres showed a marked increase in ILSS (Section 4.1) compared to composites made from untreated, corona treated or oxygen plasma treated fibres. These results are summarized in Table 4 and show that, in contrast to the results of Ward *et al.*[69] discussed earlier, there is only a marginal improvement in interfacial adhesion when the UHMPE fibre is treated with an oxygen plasma.

Electron microscope examination of the Spectra fibres showed very little evidence for the micropitting of the surface on the scale reported from the melt-spun fibres, so the component of the interfacial shear strength due to mechanical keying, τ_M, is negligible for these fibres under the treatment conditions reported.

The oxygen plasma treatment of Spectra did produce chemical modification of the surface and the XPS deconvoluted XPS spectrum in the carbon 1s region showed the presence of C—O, C=O and O—C=O groups (Table 5). Plasma and corona treatment of polyethylene film (Section 6) was reported to produce epoxide groups on the surface, but the lack of high bond strength in either

Table 5. Oxidation product distribution by XPS analysis of surface-modified ultra-high-modulus polyethylene fibres (Spectra 1000)[12,13]

	Commercial corona	Oxygen plasma		Free radical oxidation	
		15s	20 min	60 min(A)	180 min(B)
C—O	18.7	11.2	12.2	5.7	8.7
C=O	2.3	2.9	2.7	1.5	2.7
O—C=O	0.5	1.5	4.0	0	0

Concentration is expressed as % atomic weight from the area of the C 1s peak.

corona or plasma treated Spectra fibres with the epoxy resin suggests they were not present. The marginal improvement in ILSS compared to the untreated fibres suggests that wetout of the fibres was good even before polar groups were introduced. This supports the observations of Nardin and Ward[33] that the ultra-oriented fibre surface has a zero contact angle with an epoxy resin.

The only negative feature of the increased interfacial adhesion of ammonia plasma treated Spectra UHMPE composites with epoxy resins has been a reported decrease in the transverse ballistic impact properties of 10.5, 13.3 and 21.7% for plasma treatment times of 1, 2 and 10 minutes, respectively.[73] This is consistent with the observations on the loss of impact strength of strongly bonded composites from melt-spun fibre which was attributed to inhibition of debonding as an energy absorption mode. In a study of the comparative ballistic performance of Dyneema UHMPE– and Kevlar–epoxy composites it was noted that the initial fracture energy of the UHMPE–epoxy was less than one-third that of the Kevlar–epoxy.[74] Multiple delaminations in the UHMPE were required to achieve optimum impact performance and the increase in ILSS following surface modification will inhibit this process of energy absorption.

However, it was found[13] that optimum ILSS enhancement in Spectra UHMPE–epoxy was obtained after 1 min of ammonia plasma treatment of Spectra fibres (Figure 9), so the loss in ballistic impact performance may be minimal for a significant gain in ILSS. Under the mild conditions employed, neither the oxygen plasma treatment nor the ammonia plasma produced a significant loss in tensile properties of the fibre. This is consistent with the absence of surface pitting.

The ammonia plasma treatment producing >2 amine grafts/nm^2 was reported to achieve the maximum interfacial bond possible with Spectra fibres as SEM studies of the composite failure surface showed extensive fibre fibrillation and internal shear failure. It was noted in Section 2 that ultra-oriented polyethylene had a very low transverse tensile strength and an estimate of its value may be made from studies of other ultra-oriented fibres such as Kevlar, in which the transverse tensile strength of epoxy composites was 0.9% of the longitudinal strength.[16] The failure occurred by splitting of the fibres which reflects the low intermolecular forces between the highly oriented aramid chains. An estimate of the transverse tensile strength of UHMPE would thus be ~ 20 MPa, based on a longitudinal tensile strength of 2.3 GPa. This is most probably an upper limit since, in the absence of entanglements, the transverse properties of UHMPE will be controlled by dispersion forces between the closely packed oriented chains.

The interfacial adhesion in the plasma treated Spectra will be dominated by the chemical bond formed between the amine group and the epoxy, since the C—N bond energy is 286 kJ mol^{-1} compared to ~ 50 kJ mol^{-1} for a hydrogen bond. Only a fraction of the available NH$_2$ groups formed on the surface would have to form a bond with the epoxy resin to exceed the transverse tensile strength of the fibre. The limit to the effective interfacial adhesion is thus not the surface reactivity of the Spectra fibre but rather its transverse tensile strength. A similar conclusion was reached by the studies of Ward on both the melt-spun and gel-spun fibres,[63,69] and there are several strategies being researched to improve this property, such as radiation cross-linking the fibre.[75] The driving force in much of this research has been the improvement in the creep properties of melt-spun fibre (which has a low molecular weight), but the technique may also be of value in gel-spun fibres.

A particularly interesting study in this context is the investigation of the interfacial shear strength of ion beam modified Spectra 1000 fibres in epoxy composites.[76] The fibres were implanted with ions ($N^+/30$ keV; $Ar^+/75$ keV and $Ti^+/100$ keV) to an optimum dose of 0.5–1.0 × 10^{15} ions/cm^2. The interfacial adhesion was measured by the single droplet pullout technique (Section 4.3) using a bisphenol A–epoxy resin cured with *m*-phenylene diamine. The interfacial shear strength, τ, increased from 2.46 MPa for the control sample to 7.15 MPa for the best treatments ($Ti^+/100$ keV or $Ar^+/75$ keV). There was no decrease in the tensile strength of the fibre. The authors did not propose a mechanism for the increase in τ other than to suggest that a weak boundary layer may be removed. The treatment did not pit the surface and, while XPS analysis was not performed, it is unlikely that oxidation of the surface would have occurred as the samples were handled in an inert atmosphere following implantation *in vacuo*. The authors noted there was TEM evidence for a 200 nm thick ion beam modified zone which was deeper than the projected range of the ions. It is possible that the radiation chemistry of polyethylene following ion bombardment has resulted in cross-linking of the fibre to this depth. This may have increased the transverse tensile strength of the fibre in the outer zone and thus increased the shear stress at failure. Unfortunately, an analysis of the failure surfaces was not presented. It was noted earlier (Section 6) that all plasma treatments produced cross-linking to depths greater than 1 μm in unoriented polyethylene due to the high-energy UV radiation from the gas discharge. However, the cross-linking which may have occurred in oxygen plasma treated Spectra failed to produce the increase in ILSS that was observed from an ammonia plasma treatment.[13]

While most surface modifications of UHMPE fibres have been focused on achieving a high bond strength with epoxy resins, there is considerable interest in composites with unsaturated polyester and vinyl ester resins. In contrast to epoxy resins which cure by a ring opening polymerization, these resins contain unsaturated groups in-chain which are cross-linked by the free radical initiated polymerization of styrene. It would therefore be expected that if direct chemical bonding of the matrix to the fibre was to be achieved, then a different surface treatment from an epoxy resin would be required. In a study of the mechanical properties of composites prepared from oxygen plasma treated melt-spun fibres, Ladizesky and Ward[65] showed that there was no effect of changing to a styrene cross-linked polyester resin on the composite ILSS. This was consistent with mechanical keying of the resin to the surface pits on the treated fibre being the dominant mechanism for enhancement of adhesion.

In a recent study, George and Willis[12] investigated adhesion enhancement of Spectra yarns to unsaturated polyester resins by controlled oxidation of the surface. They suggested that hydroperoxide groups produced on the poly-ethylene surface should act as initiators for the polymerization of styrene during the cross-linking reaction:

$$ROOH \xrightarrow{\text{Co}^{II}} RO\bullet + \bullet OH$$

$$RO\bullet + CH{=}CH_2 \longrightarrow ROCH_2{-}\overset{\bullet}{C}H$$

where R is the backbone of the polyethylene chain. The Co^{II} redox catalyst is present as part of the normal initiation system for the unsaturated polyester resin. The hydroperoxides on the polyethylene were produced by auto-oxida-tion using a free radical initiator and were found to be stable at room temperature. Since XPS analysis of peroxides is difficult and non-specific, chemiluminescence (CL) was used for determination of their relative concentra-tion (Section 5.1.4). The change in concentration, determined in this way, with time of auto-oxidation is shown in Figure 10 and was used to determine the optimum treatment time. The auto-oxidation also produced secondary oxida-tion products, as may be seen from Table 5, which shows atomic concentration data from the XPS spectra of samples taken at points A and B of Figure 10.

The interfacial adhesion was determined by the fibre bundle pullout technique (Section 4.3) and compared with samples subjected to corona discharge and oxygen plasma treatment.[12,13] The combined results are shown in Figure 11 as pullout force as a function of embedded length of yarn and it is seen, by comparing the slopes of these lines, that while the hydroperoxidized fibres have a higher adhesion than the commercial corona discharge treated material, the oxygen plasma treatment was the most efficient of all treatments. XPS analysis showed the corona discharge treated sample had the highest surface oxygen concentration (a curve for this material was shown in Figure 4, Section 5.1.3) but this, while ensuring a strong component from polar interactions, did not achieve the bond strength of the hydroperoxidized sample. This provided strong support for the covalent bonding through polystyrene grafting as given in the earlier mechanism. Further evidence for covalent bonding was given from SEM examination of both the hydroperoxidized fibre and a typical resin groove (Figure 12). The right edge of the fibre shows the presence of a resin fragment as well as longitudinal striations, suggesting that some of the outer skin has

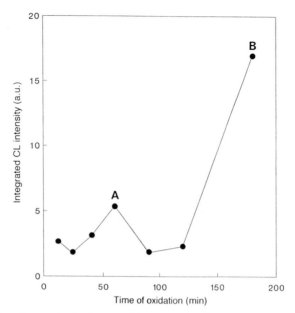

Figure 10. The change in hydroperoxide concentration as measured by integrated chemiluminescence intensity (see Figure 6) with time of auto-oxidation. The concentration of secondary oxidation products at points A and B is given in Table 5. (Reproduced from George and Willis[12] by permission of Carfax Publishing Company)

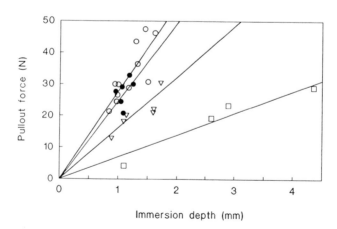

Figure 11. Pullout force for multifilament Spectra 1000 yarn from a disc of polyester resin as a function of immersion depth for: □ untreated; ▽ corona treated; ● hydroperoxidized, 60 min; and ○ 30 s oxygen plasma treated material. (Reproduced from Chappell et al.[13] by permission of Commonwealth of Australia)

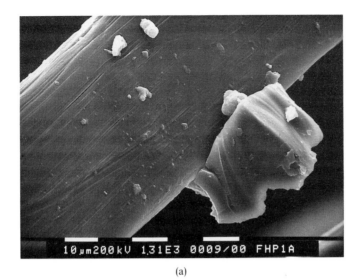

(a)

(b)

Figure 12. (a) Electron micrograph of hydroperoxidized Spectra fibre after pullout from polyester resin, showing a bonded resin fragment and longitudinal striations indicating stripping of the fibre. (b) Detail of a resin groove after pullout of a hydroperoxidized fibre, showing embedded fibrillar material due to shear failure within the fibre skin. (Reproduced from George and Willis[12] by permission of Carfax Publishing Company)

been stripped away. The resin groove (at higher magnification) clearly shows fibrillar material which has been sheared from the fibre at pullout. It was concluded that the maximum interfacial adhesion had been achieved after hydroperoxidation, given by peak A in Figure 10, and further treatment produced secondary oxidation products (notably carbonyl groups) which produced no further enhancement of bonding. In consideration of this, it is interesting that plasma treatment produced a higher bond strength. Chemiluminescence analysis has shown that a plasma treated sample of Spectra, while having a slightly different profile from a hydroperoxidized sample, contained peroxidic species able to initiate the styrene cross-linking reaction. Covalent bonding to the unsaturated polyester resin is thus possible. Since the covalently bonded hydroperoxidized Spectra had achieved a bond strength already limited by the shear strength of the fibre itself, the increased bond strength from the plasma treated fibre suggests that the cross-linking due to the vacuum UV radiation has modified the transverse tensile properties of the fibre. Alternatively, as suggested by Allred *et al.*,[10] the cross-linking of the surface may stabilize the functional groups introduced by the oxygen or amine plasma which may otherwise reorient towards the bulk to lower the surface energy.[40]

Any further improvement in the interlaminar properties of gel-spun polyethylene fibres will clearly require development of techniques for modifying the transverse tensile strength without affecting the longitudinal tensile strength and stiffness.

ACKNOWLEDGEMENTS

The research into surface modification of fibres involves collaboration between the Polymer Materials and Radiation Group in this Department (Ms Heather Willis and Dr Richard Ward) and Materials Research Laboratory, Defence Science and Technology Organization, Melbourne (Dr James Brown and Dr Philip Chappell). Particular thanks are extended to Drs Brown and Chappell for providing copies of their manuscripts prior to publication.

REFERENCES

1. Ashbee, K., *Fundamental Principles of Fiber Reinforced Composites*, Technomic, Pennsylvania (1989).
2. Hale, D. K., and Kelly, A., *Ann. Rev. Mater. Sci.*, **2**, 405 (1972).
3. Caldwell, D. L., Interfacial analysis, in S. M. Lee (ed.), *International Encyclopedia of Composites*, Vol. 2, VCH Publishers, New York (1990) p. 361.
4. Drzal, L. T., *Adv. Polym. Sci.*, **75**; K. Dusek (ed.), *Epoxy Resins and Composites*, Vol. II, Springer-Verlag, Berlin (1986), p. 1.
5. Bascom, W. D., Interphase in fiber reinforced composites, in S. M. Lee (ed.), *International Encyclopedia of Composites*, Vol. 2, VCH Publishers, New York (1990), p. 411.

6. Kalnin, I. L., and Jager, H., in E. Fitzer (ed.), *Carbon Fibres and Their Composites*, Springer-Verlag, Berlin (1985), p. 62.
7. Hattery, G. R., and Hillman, M. E. D., in E. Baer and A. Moet (eds.), *High Performance Polymers—Structures, Properties, Composites, Fibers*, Hanser Publishers, Munich (1991), p. 255.
8. Kirschbaum, R., and van Dingenen, J. L. J., in P. Lemstra and L. A. Kleintjens (eds.), *Integration of Fundamental Polymer Science and Technology*, Vol. 3, Elsevier Applied Science, London (1989), p. 178.
9. Kalantar, K., and Drzal, L. T., *J. Mater. Sci.*, **25**, 4186 (1990).
10. Allred, R. E., Merrill, E. W., and Roylance, D. K., in H. Ishida (ed.), *Molecular Characterization of Composite Interfaces*, Plenum Press, New York (1984), p. 333.
11. Ward, I. M., and Ladizesky, N. H., *Pure Appl. Chem.*, **57**, 1641 (1985).
12. George, G. A., and Willis, H. A., *High Performance Polym.*, **1**, 335 (1989).
13. Chappell, P. J. C., Brown, J. R., George, G. A., and Willis, H. A., *Surf. Interf. Anal.*, **17**, 143 (1991).
14. Dobb, M. G., Johnson, D. J., and Saville, B. P., *Polymer*, **22**, 960 (1981).
15. van der Zwag, S., Picken, S. J., and van Sluijs, C. P., in P. J. Lemstra and L. A. Kleintjens (eds.), *Integration of Fundamental Polymer Science and Technology*, Vol. 3, Elsevier Applied Science, London (1989), p. 199.
16. Clements, L. L., and Chiao, T. T., *Composites*, **87** (April 1977).
17. Stachurski, Z. H., in *Engineering Science of Polymeric Materials*, Polymer Division RACI, Melbourne (1987), p. 189.
18. Plueddemann, E. P., *Silane Coupling Agents*, Plenum, New York (1982), p. 1.
19. Browning, C. E., Abrams, F. L., and Whitney, J. M., ASTM STP797, American Society for Testing and Materials, Philadelphia (1983), p. 54.
20. Smiley, A. J., and Pipes, R. B., *J. Composite Mater.*, **21**, 670 (1987).
21. Day, R. J., Robinson, I. M., Zakikhani, M., and Young, R. J., *Polymer*, **28**, 1833 (1987).
22. Melanitis, N., Tetlow, P., Galiotis, G., and Davies, C. K. L., in F. R. Jones (ed.), *Interfacial Phenomena in Composite Materials '89*, Butterworths, London (1989), p. 97.
23. Day, R. J., Zakikhani, M., Ang, P. P., and Young, R. J., in F. R. Jones (ed.), *Interfacial Phenomena in Composite Materials '89*, Butterworths, London (1989), p. 121.
24. Jahankhani, H., Vlattas, C., and Galiotis, C., in F. R. Jones (ed.), *Interfacial Phenomena in Composite Materials '89*, Butterworths, London (1989), p. 125.
25. Piggott, M. R., *Comp. Sci. Technol.*, **30**, 295 (1987).
26. Pitkethly, M. J., and Noble, J. B., in F. R. Jones (ed.), *Interfacial Phenomena in Composite Materials '89*, Butterworths, London (1989), p. 35.
27. Bandyopadhyay, S., and Murthy, P. N., *Mater. Sci. Engng*, **19**, 139 (1975).
28. Chappell, P. J. C., Williams, D. R., and George, G. A., *J. Colloid Interf. Sci.*, **134**, 385 (1990).
29. Wu, Y., and Tesoro, G. C., *J. Appl. Polym. Sci.*, **31**, 1041 (1986).
30. Penn, L. S., Byerley, T. J., and Lias, T. K., *J. Adhesion*, **23**, 163 (1987).
31. Brewis, D. M., and Briggs, D., *Polymer*, **22**, 7 (1981).
32. Caroll, B. J., *J. Colloid Interf. Sci.*, **57**, 488 (1976).
33. Nardin, M., and Ward, I. M., *Mater. Sci. Technol.*, **3**, 814 (1987).
34. Dorris, G. M., and Gray, D. G., *J. Colloid Interf. Sci.*, **71**, 93 (1979).
35. Briggs, D., *Polymer*, **25**, 1379 (1984).
36. Vickerman, J. C., in J. M. Walls (ed.), *Methods of Surface Analysis—Techniques and Applications*, Cambridge University Press (1989), Ch. 6.
37. Garbassi, F., Occhiello, E., Bastioli, C., and Romano, G., *J. Colloid Interf. Sci.*, **119**, 258 (1987).

38. Briggs, D., and Seah, M. P., (eds.), *Practical Surface Analysis*, John Wiley, New York (1987).
39. Dilks, A., in J. V. Dawkins (ed.), *Developments in Polymer Characterization*, Vol. 2, Applied Science UK (1980), p. 145.
40. Gillberg, G., *J. Adhesion*, **21**, 129 (1987).
41. Batich, C. D., *Appl. Surf. Sci.*, **32**, 57 (1988).
42. Carlsson, D. J., Brusseau, R., Zhang, C., and Wiles, D. M., *Polym. Deg. Stab.*, **17**, 308 (1987).
43. Everhart, D. S., and Reilly, C. N., *Surf. Interf. Anal.*, **3**, 126 (1981).
44. Holmes, S., and Schwartz, P., *Comp. Sci. Technol.*, **38**, 1 (1990).
45. Zlatkevich, L., *Radiothermoluminescence and Transitions in Polymers*, Springer-Verlag, New York (1987).
46. Boustead, I., *J. Polym. Sci.*, **A2**(8), 143 (1970).
47. Fisher, W. K., *J. Indust. Irradiation Technol.*, **3**, 167 (1985).
48. Billingham, N. C., Then, E. T. H., and Gijsman, P. J., *Polym. Deg. Stab.*, **34**, 263 (1991).
49. George, G. A., and Ghaemy, M., *Polym. Deg. Stab.*, **34**, 37 (1991).
50. George, G. A., and Ghaemy, M., *Polym. Deg. Stab.*, **33**, 411 (1991).
51. Blythe, A. R., Briggs, D., Kendall, C. R., Rance, D. G., and Zichy, V. J. I., *Polymer*, **19**, 1273 (1978).
52. Everhart, D. S., and Reilly, C. N., *Anal. Chem.*, **53**, 665 (1981).
53. Gerenser, L. J., Elman, J. F., Mason, M. G., and Pochan, J. M., *Polymer*, **26**, 1162 (1985).
54. Gerenser, L. J., *J. Adhesion Sci. Technol.*, **1**, 303 (1987).
55. Eriksson, J. C., Golander, C.-G., Baszkin, A., and Ter-Minassian-Saraga, L., *J. Colloid Interf. Sci.*, **100**, 381 (1984).
56. Hansen, R. H., and Schonhorn, H., *J. Polym. Sci.*, **134**, 203 (1966).
57. Schonhorn, H., and Ryan, F. W., *J. Polym. Sci.*, **A2**(6), 231 (1968).
58. Briggs, D., Brewis, D. M., and Konieczko, M. B., *J. Mater. Sci.*, **12**, 429 (1977).
59. Iwata, H., Kishida, A., Suzuki, M., Hata, Y., and Ikada, Y., *J. Polym. Sci. A (Polym. Chem.)*, **26**, 3309 (1988).
60. Foerch, R., and McIntyre, N. S., *J. Polym. Sci. A (Polym. Chem.)*, **28**, 193 (1990).
61. Rapoport, N. Ya., Goniashvili, A. Sh., Akutin, M. S., and Miller, V. B., *Polym. Sci. USSR*, **21**, 2286 (1979).
62. Rapoport, N. Ya., Goniashvili, A. Sh., Akutin, M. S., Shibryayeva, L. S., Ponomareva, Ye, L., and Miller, V. B., *Polym. Sci. USSR*, **23**, 439 (1981).
63. Ladizesky, N. H., and Ward, I. M., *J. Mater. Sci.*, **18**, 533 (1983).
64. Ladizesky, N. H., and Ward, I. M., *Comp. Sci. Technol.*, **26**, 129 (1986).
65. Ladizesky, N. H., and Ward, I. M., *Comp. Sci. Technol.*, **26**, 169 (1986).
66. Ladizesky, N. H., and Ward, I. M., *Comp. Sci. Technol.* **26**, 199 (1986).
67. Rozenberg, B. A., *Adv. Polym. Sci.*, **75**; K. Dusek (ed.), *Epoxy Resins and Composites*, Vol. II, Springer-Verlag, Berlin (1986), p. 113.
68. Morgan, R. J., and Mones, E. T., *J. Appl. Polym. Sci.*, **33**, 999 (1987).
69. Tissington, B., Pollard, G., and Ward, I. M., *J. Mater. Sci.*, **26**, 82 (1991).
70. Kaplan, S. L., Rose, P. W., Nguyen, H. X., and Chang, H. W., *Proceedings of the 33rd International SAMPE Symposium*, SAMPE, Corina (1988), p. 551.
71. Nguyen, H. X., Riahi, G., Wood, G., and Poursartip, A., *Proceedings of the 33rd International SAMPE Symposium*, SAMPE, Corina (1988), p. 1721.
72. Wagner, H. D., and Cohn, D., *Biomaterials*, **10**, 139 (1989).
73. Brown, J. R., Chappell, P. J. C., and Mathys, Z., *J. Mater. Sci.*, **26**, 472 (1991).

74. Scholle, K., in P. J. Lemstra and L. A. Kleintjens, (eds.), *Integration of Fundamental Polymer Science and Technology*, Vol. 3, Elsevier Applied Science (1989), p. 212.
75. Ward, I. M., in P. J. Lemstra and L. A. Kleintjens (eds.), *Integration of Fundamental Polymer Science and Technology*, Vol. 1, Elsevier Applied Science (1986), p. 634.
76. Ozollo, A., Grummon, D. S., Drzal, L. T., Kalantar, J., Loh, I.-H., and Moody, R. A., *Mater. Res. Soc. Symp. Proc.*, **153**, 217 (1989).

8

SSIMS—An Emerging Technique for the Surface Chemical Analysis of Polymeric Biomaterials

M. C. Davies

Department of Pharmaceutical Sciences,
University of Nottingham

1 INTRODUCTION

The surface chemistry of polymeric biomaterials has a major impact on their fate *in vivo*. The surface of any polymeric device placed within the biological milieu will elicit a rapid and complex series of biological events as part of the body's defence mechanism against foreign materials. Within seconds, proteins will adsorb at the polymer interface and the composition and extent of this response will largely determine the long-term fate of the biomaterial. In this light, it is not surprising that there has been considerable interest within biomedical science in examining the relationship between the chemical and physical nature of the interfacial structure and the surface energetics exhibited.[1-2] Similarly, the complex recognition processes responsible for the nature of cellular and protein adhesion[3] to these polymeric surfaces are also the subject of intense study.

With this in mind, there have been significant advances in biomaterials design over the last decade which focus on two specific and very different themes: firstly, the development of surfaces which exhibit good biocompatability *in vivo* and elicit minimal protein and cellular adhesion and secondly, the development of surfaces that promote highly selective interactions such as biosensors,[4,5] immunolattices[6,7] and targeting drug delivery systems.[8,9] Only a few polymeric materials have the appropriate surface chemical structure to exhibit good biocompatibility. In general, the requirements in the polymeric formulation for optimal bulk properties in terms of (say) mechanical strength or permeability frequently produce a poorly tolerated surface *in vivo*.

Thus, a significant effort has been focused in biomaterials science on the development of polymeric surfaces designed to exhibit a specific interfacial chemistry. A diverse range of surface modification procedures[10] commonly

Polymer Surfaces and Interfaces II
Edited by W. J. Feast, H. S. Munro and R. W. Richards
© 1993 John Wiley & Sons Ltd

employed within polymer and materials science have been exploited. Thin film coating techniques such as Langmuir–Blodgett technology[11] have been used to deposit highly organized mono- and multilayers at interfaces. Plasma modification and polymerization[12–13] have been used to improve the surface energetics of the biomaterials. One specific area of intense interest lies in the field of development of sophisticated organic syntheses at surfaces that are designed to construct a specific molecular architecture at the polymer inter-face.[11] To this end, biomolecules such as lipids have been immobilized to polymer surfaces to mimic cellular membranes and hence elicit good biocompat-ibility.[14–16] In contrast, proteins and carbohydrates have been covalently bound to surfaces to promote specific molecular interactions.[17] Polymers such as poly(ethylene oxide) have been grafted to surfaces to produce a hydrophilic steric barrier which reduces protein adsorption significantly.[18–20]

However exquisite the surface engineering, the surface composition of these polymeric biomaterials will also be defined by the processing, handling and analysis conditions. The presence of contaminants such as poly(dimethylsil-oxane), or PDMS, and the migration of one or more component to the surface of copolymers will elicit a change in the predicted surface chemistry. Care must also be taken in data interpretation from the surface analysis of hydrated systems by methods employing high vacuum, as significant surface reorientation is likely to occur.

In all these applications, a clear definition of the surface chemical structure of the polymer will allow some understanding of the nature and extent of the surface engineering and also provide some insight into the relationship between the interfacial structure and biointeractions observed.[21] Such an approach requires an interdisciplinary effort and needs to address a number of key issues. These include the molecular structure and spatial arrangement of functional groups, their mobility and reorientation, and their influence on the surface chemistry, and the analysis of impurities and additives.[21]

Such questions pose considerable fundamental analytical problems but the exploitation of sophisticated surface analytical techniques has been recognized as a fruitful approach to resolve some of the above issues.[21] To achieve a detailed analysis of biomaterial surface structure, a number of key features are required: quantitative elemental and molecular information, an insight into the lateral spatial arrangement of functional groups and molecules, a minimum sampling depth and low damage during analysis. A range of advanced surface analysis techniques have emerged over the last twenty years for the characteriza-tion of polymer surfaces which in part, but rarely wholly, match these criteria; these are reviewed in detail elsewhere.[22–30] In particular, X-ray photoelectron spectroscopy (XPS)[23–24] has been successfully applied to a diverse range of biomaterials largely through the work of Ratner, Andrade and others.

In this chapter, the use of secondary ion mass spectrometry (SIMS) will be highlighted as a complimentary tool to XPS for the surface chemical structure of polymeric biomaterials. The theoretical and experimental aspects of SIMS and XPS are discussed in detail elsewhere.[22,24-27,31] This article will concentrate primarily on how these techniques provide valuable information in the characterization of polymeric biomaterial surfaces. While this review will concentrate on the application of SIMS in this field, wherever possible the complimentary XPS analysis will be described.

2 SIMS—THEORY AND INSTRUMENTATION

The principles and operation of SIMS instrumentation have been described in detail[31] but the salient points are summarized here for the reader. In the SIMS process (see Figure 1), a primary noble gas atom or ion (e.g. Ar^+, Xe^+, Xe^0 and Ar^0) beam bombards the sample in UHV, penetrating to a depth of

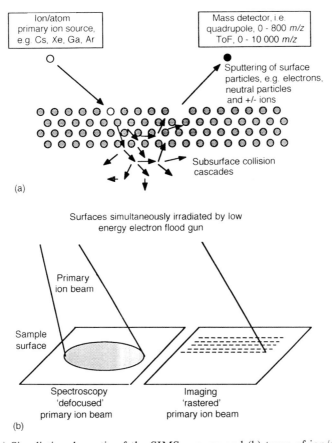

(a)

(b)

Figure 1. (a) Simplistic schematic of the SIMS process and (b) types of ion/atom gun for SIMS (reproduced by kind permission of CRC Press)

30–100 Å. The kinetic energy of the particle is thought to dissipate via a collision cascade process which causes the emission of electrons, neutral species and secondary ions (5% of sputtered material) remote from the initial primary ion impact. The secondary ion yields vary with polymer surface composition. The noted influence of the surface composition on secondary ion yields, known as 'matrix effects', frequently obviates the possibility of quantitative SIMS information.[32]

The positively and negatively charged ions are analysed in terms of their mass-to-charge ratio by a quadrupole or a time-of-flight (ToF) mass spectrometer, yielding a positive and negative secondary ion mass spectrum. The majority of polymer SIMS analyses to date has utilized quadrupole mass spectrometers with typical m/z ranges of 0–800. More recently, ToF SIMS instruments[33] with improved transmission characteristics, good mass discrimination and high working mass ranges ($m/z = 0$–10 000 Da) have been developed.

Significant care must be taken in the analysis conditions for SIMS spectra of polymers. Polymeric materials are very susceptible to damage by ionizing radiation and the maximum ion/atom dose to ensure that negligible sputtering of the upper monolayer occurs during the SIMS experiment[34–35] (referred to as static SIMS or SSIMS) may be as low as 10^{13} particles/cm². The sampling depth for the SSIMS analysis of polymers[36] under these 'static' conditions is of the order of 10–12 Å. Since the irradiation of the sample may induce sample charging, which will suppress secondary ion yields, the optimum conditions for surface potential control by the simultaneous irradiation with a flood of low-energy electrons have been devised.[37–38]

3 SSIMS AND XPS OF POLYMER SURFACES

Since the first reports of SSIMS analyses of polymers, a wide range of polymer classes has been analysed (see Table 1) and a number of libraries of SSIMS spectra of polymers has been published. In contrast to the simplistic assignment of binding energy shifts within the core level XPS peaks, SSIMS spectra may appear complex and difficult to interpret. In reality, the conventional and simple rules of mass spectrometry[39] may be applied to SSIMS, e.g. positive ions within SSIMS spectra may be formed by the loss of a bonding or, to a lesser extent, non-bonding electron to form 'even electron' or 'odd electron' ions (also known as radical cations), respectively.[39–40]

There has been some interest in understanding the mechanisms of polymer fragmentation in ion formation. In these studies, SSIMS spectra have been correlated with data from other techniques such as pyrolysis mass spectrometry[41–42] and other processes where significant information is available on ion formation and polymer fragmentation.[43] Significant effort is also being focused on the development of methodologies to assist in the elucidation of the

Table 1. SSIMS analysis of polymers and copolymers of biomedical significance

Polymer	Biomedical use	Reference

Poly(esters)
  Poly(glycolide)
  Poly(lactide)
  Poly(hydroxybutyrate)
  Poly(valerate)
  Poly(caprolactone)
	Biodegradable polymers
	for drug delivery
		42,51,59

Poly(orthoesters) see section 4.3	Biodegradable polymers for drug delivery	54

Poly(anhydrides)
  Poly(sebacic anhydride)
  etc.
	Biodegradable polymers
	for drug delivery
		99

Poly(alkylacrylates)
  methyl, ethyl and n-butyl
Poly(alkylmethacrylates)
  methyl- to s-butyl, and cyclohexyl
Poly(hydroxyalkylmethacrylates)
  hydroxyethyl- and
  hydroxypropyl-
Poly(methacrylic acid)
Poly(methacrylic anhydride)
	Hydrogels, bone cement	38,43,105
		57,58

Nylons
  Nylons 4, 6, 7, 8, 11 and 12
  Nylons 46, 66 and 610
		41

Poly(urethanes) see section 4.3	Prostheses, catheters, heart valves	36,55
Poly(carbonates)		66,106

Poly(alkylsiloxanes)
  dimethyl-
  phenyl-
	Catheters, membranes,
	rate controlling matrices in
	drug delivery
		26,107

Poly(ethers)
  Poly(ethylene glycol)
  Poly(propylene glycol)
  Poly(tetramethylene glycol)
	Hydrogels, soft segments in
	polyurethanes, protein resistant
	surfaces, steric barriers for
	colloids
		55

Cellulose ethers and esters
  Methyl-
  Ethyl-
  Hydroxypropyl-
  Hydroxpropylmethyl-
  Hydroxpropylmethylsuccinate-
  Hydroxypropylmethylphthalate-
	Membranes, matrices for drug
	delivery
		100,88

Hydrocarbon polymers
  Poly(styrene)
  Poly(ethylene)
  Poly(propylene)
  Poly(butadiene)
  Poly(isoprene)
	Tissue culture plates,
	general applications
		101–103

Miscellaneous Polymers
  Poly(vinyl alcohol)
  Poly(ethyleneterephthalate)
	Hydrogels	45
		102,104

precise structure of some of the prominent ions within SSIMS spectra. The labelling of one or more portions of the monomer unit with an isotopic element has yielded valuable information on ion structures in SSIMS spectra of poly(methyl methacrylate),[44] poly(vinyl alcohol)[45] and a plasma polymer.[46] Similarly, gas derivatization procedures employed in XPS have also been successfully employed.[47] The recent advent of tandem mass spectrometry in SSIMS analyses where a daughter ion mass spectrum is generated from a dominant primary ion within the mass spectrum has greatly assisted in the assignments of certain ion structures within a number of polymer classes including hydrocarbon polymers.[48–50]

For the purpose of this chapter, the application of SSIMS and XPS to polymeric biomaterials will be illustrated by two examples: a polymer possessing a simple monomer structure and a complex copolymer. In both these cases, the SSIMS spectra shown are typical of the growing body of SIMS literature on polymers where the observed fragmentation patterns may be interpreted in terms of structures arising from the intact monomer unit.[25–28]

Figures 2 and 3 show the positive and negative ion SSIMS spectrum of a biodegradable polymer, poly(lactic acid) or PLA,[42] which is from a class of polyesters that is currently the subject of intense interest as vehicles for advanced drug delivery systems. Both SSIMS spectra are dominated by ions corresponding to structures which may be assigned to $nM \pm H$, where M and n are the mass and the number of the monomer repeat units, at m/z 71/73, 143/145, 215/217 and 287/289, for $n = 1$–4, respectively; e.g.

PLA polymer structure

$$\left[\begin{array}{c} CH_3 \quad O \\ | \qquad \| \\ CH-C-O \end{array}\right]_n$$

Major ions

$$^+HC\overset{CH_3}{\underset{|}{}}\overset{O}{\underset{\|}{-C}}-OH \qquad H_2C\overset{CH_3}{\underset{|}{}}\overset{O}{\underset{\|}{-C}}-O^-$$

$(M+H)^+$ $(M+H)^-$

$$^+HC\overset{CH_3}{\underset{|}{}}\overset{O}{\underset{\|}{-C}}-O\left[\begin{array}{c}CH_3 \quad O \\ | \qquad \| \\ CH-C-O\end{array}\right]_{n-1}H$$

$(nM+H)^+$

Further ions within the SIMS spectra may be assigned to structures corresponding to $(nM + O \pm H)^+$, $(nM - O \pm H)^-$ and $(nM - O)^+$. Subtle changes in the chemical structure of the monomer unit, e.g. as for poly(glycolic acid)[42] and poly(caprolactone),[51] produce the corresponding shift in the mass of the

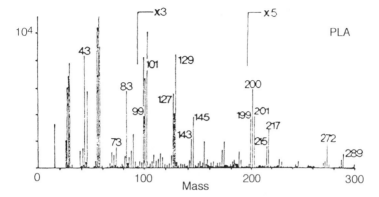

Figure 2. Positive ion SIMS spectrum of PLA[42]

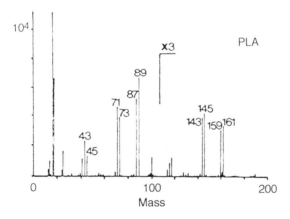

Figure 3. Negative ion SIMS spectrum of PLA[42]

molecular ions within the SSIMS spectra, allowing the ready differentiation of structurally similar polymers. Thus, one is able to provide a detailed molecular picture of a homopolymer by the simple application of mass spectrometry rules to the interpretation of the SSIMS spectrum. In general, the production of $M \pm H$ and related ions is observed as a general feature of the SSIMS analyses of many polymers. However, for reasons noted above,[32] it is not possible to directly quantify the level of functional groups or elements from the SIMS analysis.

In contrast, the XPS data[42] are able to qualitatively confirm the elemental purity of the surface and provide quantitative elemental analysis which for PLA (the carbon-to-oxygen ratio is 1.49:1) has been shown to be in good agreement with theoretically determined values (1.5:1). The chemical shifts employed

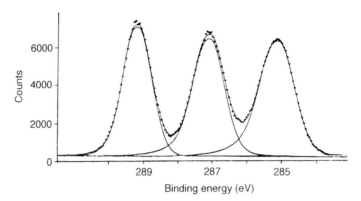

Figure 4. Carbon 1s spectrum of PLA produced with high-resolution XPS instrumentation (courtesy of Drs G. Beamson and D. Briggs, ICI)

within the carbon 1s envelope fits are also in very close agreement with those determined for $\underline{C}$—H, $\underline{C}$—O and O—$\underline{C}$=O bonding environments.[42] Recently, new instrumentation possessing an incident monochromated beam with a greater flux than conventional machines and with greatly increased transmission through the use of larger electron analysers and detectors has allowed very high energy resolution XPS spectra to be obtained.[52–53] This higher resolution is illustrated in Figure 4 for PLA. The systematic work of Beamson and Briggs with this instrument suggests that it will significantly advance our understanding of both fundamental and applied aspects of the XPS analysis of polymer surface chemistry. Despite the quantitative chemical state information available from XPS analysis, it is naturally not possible to determine the molecular arrangement of the functional groups within a monomer unit irrespective of the resolution obtained. Hence, the symbiotic nature of the combined application of SIMS and XPS for polymer surface chemical analysis is evident from the analysis of even this simple polymer system.

The potential of XPS and SSIMS for the analysis of the surface structure of complex copolymer systems may be illustrated by the analysis of a range of poly(orthoesters)[54] which are known to exhibit pH-controlled surface degradation and have considerable potential for controlled drug delivery. The orthoester copolymers may be prepared with a flexible diol (1,6-hexanediol[HD]) and a rigid diol (*trans*-cyclohexanedimethanol [CDM]) as shown below:

where

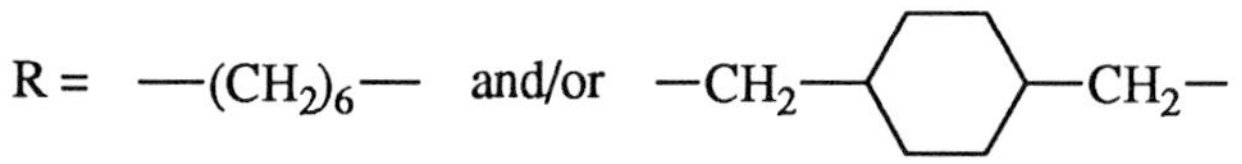

$$R = \quad -(CH_2)_6- \quad \text{and/or} \quad -CH_2-\!\!\langle\bigcirc\rangle\!\!-CH_2-$$

flexible diol   rigid diol

It is possible to fit four chemical states to the carbon 1s envelope,[54] i.e. $\underline{C}-C/\underline{C}-H$, $\underline{C}-O$, $\underline{C}O_3$ and $\underline{C}-CO_3$, which, together with the elemental analysis, show an excellent correlation with those anticipated from the structural formula, confirming the polymer purity. Again, these is no information of the arrangement of these bonding environments within a copolymer structure.

The positive ion SSIMS spectrum of a poly(orthoester) with the rigid diol, CDM, is shown in Figure 5. By systematic analysis of a range of poly-(orthoesters),[54] it is possible to assign the cations within the mass spectrum to structures arising from the diol, the orthoester unit and the orthoester–diol monomer unit. The intense m/z 109 ion in Figure 5 is not observed in any other orthoester–diol copolymer and is assigned to the following structure from the CDM unit with the associated primary carbonium ion:

$$^+CH_2-\!\!\langle\bigcirc\rangle\!\!=CH_2 \quad \longleftarrow \quad CH_3-\!\!\overset{+}{\langle\bigcirc\rangle}\!\!=CH_2$$

m/z 109

The series of ions at m/z 231/229, 215/213, 201, 175, 159, 143 and 127 are observed in the same relative intensity in all poly(orthoesters)[54] investigated and are thought to correspond to structures from the orthoester unit, e.g. $(M \pm H)$, $(M - O \pm H)$ and $(M - CH_2CH_3)$ at m/z 231/229, 215/213 and 201, respectively:

Possible ion assignments	*m/z*	Notation
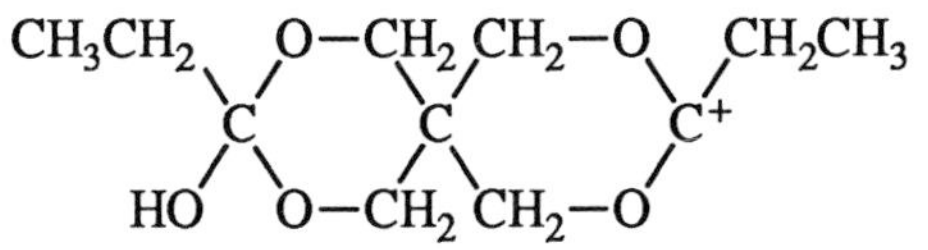	231	$(M + H)^+$
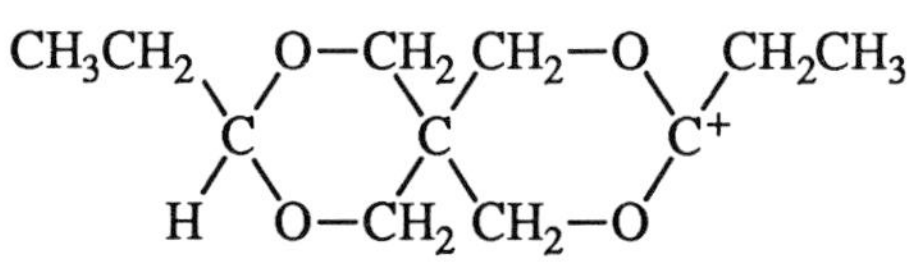	215	$(M - O + H)^+$
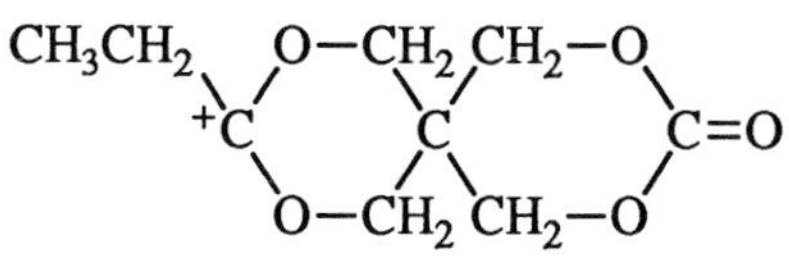	201	$(M - CH_2CH_3)^+$

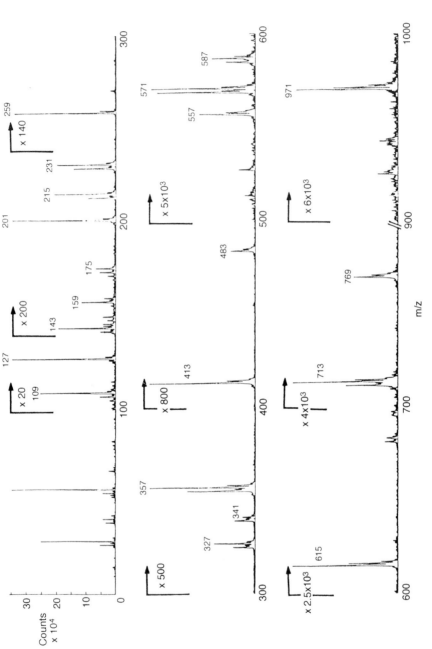

Figure 5. Positive ion ToF-SIMS spectrum (m/z 0–1000) of poly(*ortho*-ester) containing CDM diol.[54] (Reprinted with permission from Davies *et al.*[54] Copyright (1991) American Chemical Society)

The ions in the m/z 300–1000 range are derived from the intact copolymer structure, e.g. $n(M_{\text{orthoester}} + M_{\text{CDM}}) \pm H$ is observed at m/z 357/355, 713/711 and (not shown) 1069/1067 for $n = 1–3$, respectively. Ions corresponding to $n(M_{\text{orthoester}} + M_{\text{CDM}}) - O \pm H$ and $n(M_{\text{orthoester}} + M_{\text{CDM}}) - CH_2CH_3$ are also observed.[54]

When the poly(orthoester) copolymers are prepared containing both the rigid and flexible diol, both sets of ions are observed within the mass spectrum which are derived from both $(M_{\text{orthoester}} + M_{\text{HD}})$ and $(M_{\text{orthoester}} + M_{\text{CDM}})$.[54] In addition, there is a series of ions in these copolymers that is not observed in the SSIMS spectra of orthoesters containing a single diol. These cations are composed of structures containing both HD and CDM diols and the orthoester unit, and hence arise from the random sequence of the copolymer chain. Thus, it is clear that SSIMS spectra can distinguish different organic structures within a complex copolymer. Similar findings have been reported for other biomedical polymers such as poly(ether urethanes)[36] where the hard and soft segments were readily distinguished.

A particularly interesting feature of the poly(orthoester) study[54] was the comparison of the relative ion intensities of ions diagnostic of HD and CDM ratioed to the dominant ions from the poly(orthoester) structure as a function of copolymer composition (see Figure 6). There is a near linear relationship increase in the ion ratios for both CDM and HD with bulk levels. These data suggest that the peak area of the ions within the mass range reflect the nominal bulk composition in a quantitative manner. Similar findings were reported earlier in SSIMS studies on poly(urethanes)[36] where the ratio of ions for each

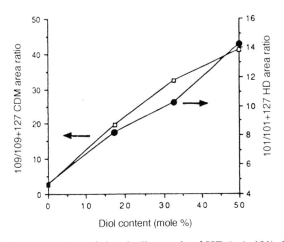

Figure 6. Plot of the area ratios of signals diagnostic of HD (m/z 101), CDM (m/z 109) and the *ortho*-ester unit (m/z 127) as a function of the bulk polymer composition. (Reprinted with permission from Davies *et al.*[54] Copyright (1991) American Chemical Society)

hard and soft segment showed an excellent correlation with the ratio of each segment derived from XPS data over a wide range of compositions. While it is recognized that the influence of matrix effects on relative ion yields usually precludes the direct quantitation of ion intensities, there is a growing body of SSIMS literature on polymers where secondary ion yields have been shown to correlate with bulk copolymer composition, e.g. nylons,[41] poly(ether urethanes),[36,55] poly(alkyl methacrylates)[56-58] and poly(esters).[59]

An interesting advance in the analysis of copolymer surface composition by SSIMS involved the sequencing of a series of ethyl methacrylate (EMA) and hydroxyethyl methacrylate (HEMA) copolymers.[56] A comparison of the relative intensity of ions characteristic of each monomer unit and linked HEMA–EMA monomer units was found to be consistent with a random sequence composition, suggesting in this case that the surface composition was the same as the bulk. The authors speculated that the deviations from the random copolymer sequence may be readily detected. As the presence of non-random or blocky sequences is likely to have a significant effect on phase separation or preferential surface segregation, and hence surface energetics and reactivity of copolymers, such studies should be an important goal within biomedical polymer surface analysis.

An interesting application of the ToF-SIMS instrument concerning the analysis of oligomer molecular weight distributions has been demonstrated by Benninghoven and his coworkers.[60-62] Such high-mass fragments lie outside the range of conventional quadropole SIMS instruments. Oligomeric distributions of up to 10 kDa have been measured which have been shown to correlate well with other analytical techniques such as gel permeation chromatography. With the increasing interest in designing soluble low molecular weight polymeric–drug conjugates for site-specific drug delivery, ToF-SIMS may have a particularly interesting role in the characterization of these systems.

4 THE ANALYSIS OF MODIFIED POLYMER SURFACES

The diverse range of surface modification procedures employed within biomedical research have been highlighted in the introduction to this paper. The change in the interfacial structure of a biomaterial device may be critical for ensuring the optimum biocompatibility profile while ensuring that the desired bulk properties remain unchanged. Such approaches to interfacial modification have been primarily examined by XPS and, to date, relatively few examples exist within the literature on the application of SIMS to the study of such complex surfaces.

4.1 Surface Modification by Plasma Treatment and Polymerization

The surface modification procedures such as plasma etching and polymerization have been examined by XPS and, to a limited extent, by SSIMS. The gradual change in surface wettability of a contact lens consisting of an alkyl acrylate, an organosilane unit with treatment time was interpreted by SIMS and XPS to arise from a progressive conversion of the surface organosiloxanes species to a surface silica phase.[63–64] The effect of plasma gas on the glow discharge treatments on a range of polymers such a poly(ethylene),[65] poly(propylene)[65] and poly(carbonate)[66–68] has also been studied by SIMS. Where the polymer is labile to the plasma treatment, the change of the homopolymer interfacial chains into a multicopolymer surface containing many functionalities, each with a distinctive fragmentation ion, may yield complex and potentially uninterpretable SIMS spectra. In such a case, XPS analyses, particularly allied with surface derivatization procedures,[69] are likely to be more valuable in elucidating the generation of different functionalities in a quantitative manner.

4.2 Surface Segregation and Preferential Orientation at Polymer Surfaces

SSIMS may have an important complementary role to XPS in the study of polymer surface segregation. An SSIMS study on polyurethanes[36,55] by Briggs and Ratner resulted in a revision of the theoretical model of phase separation exhibited by the hard and soft segments of the copolymer from XPS and other studies. While the previous analyses and models had proposed a pure soft segment overlay on top of the hard segment, the SSIMS analysis detection of both molecules within its limited sampling depth (10 Å) suggests that the surface is more a mix of hard and soft segments than originally thought. As part of this study, SSIMS was able to provide clear evidence of increasing soft segment segregation to the surface with its increasing molecular weight.[55] In similar studies on poly(glycolide)–trimethylene carbonate (TMC) copolymers, SSIMS and angular resolved XPS studies suggested surface enrichment of the TMC component, possibly due to the presence of TMC blocks within the copolymer sequence.[70]

An interesting example of preferential surface orientation occurs for the polymer end groups in the formation of polymer colloids. The negative ion SIMS spectrum of a model poly(butyl methacrylate)–latex system prepared by surfactant-free emulsion polymerization using potassium persulphate as a free radical initiator is shown in Figure 7(a). The SIMS spectrum is dominated by ion signals at m/z 64, 80 and 96 corresponding to SO_2^-, SO_3^- and SO_4^-, respectively, arising from the sulphate end groups on the PBMA chains.[73] These groups are preferentially oriented at the particle surface due to the colloid formation process and are responsible for the negative surface potential which provides an electrostatic barrier to particle aggregation and hence colloid stability. The anions attributable to the PBMA monomer unit and the butyl sidechain are observed as weak signals at m/z 55 and 85, and m/z 71/73, respectively. The relative dominance of the ions arising from the polymer end groups over those from the polymer backbone reflects the surface density of

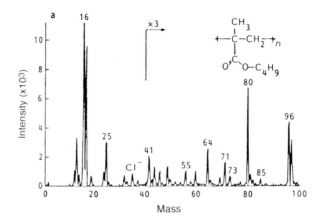

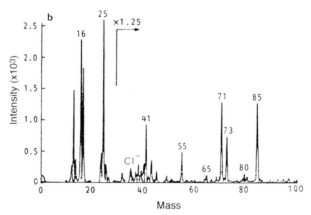

Figure 7. Negative ion SSIMS spectrum of (a) poly(butyl methacrylate) latex (m/z 0–100), and (b) poly(butyl methacrylate) latex dried and cast from chloroform (m/z 0–100). (Reprinted from Lynn *et al.*[73] by permission of Butterworths)

the sulphate end groups. This is confirmed by Figure 7(b) where the latex has been dissolved in chloroform,[73] thereby abolishing the surface orientation of the charged end groups and providing a more random distribution of the polymer molecules. In this case, the PBMA ions predominate and the sulphate groups are barely detectable. An excellent correlation has also been observed[74] for PBMA colloids prepared with different surface densities of sulphate end groups between the change in the relative intensities of the sulphate and PBMA ions in the SIMS spectra and the corresponding XPS and colloid surface charge measurements.

4.3 Surface Grafting of Biomolecules and Polymers

The development of sophisticated surface chemistry techniques for the immobilization of biomolecules to polymer and in particular colloid surfaces has many applications in the immunochemistry,[6] chromatography and drug delivery fields.[71,72,75] The nature of the surface immobilization, the surface density of the ligand and the potential denaturation of the ligand during immobilization are all critical issues. Figure 8 shows the positive ion SSIMS spectrum of a PMMA latex prepared with a thioglycoside (25% bulk composition), based on galactose with an acryloyl group in the aglycone sidechain, by surfactant-free emulsion polymerization employing potassium persulphate initiator.[76] One of the most important features of this spectrum is the presence of the dominant m/z 373 cation which corresponds to the protonated molecular ion, $(M + H)^+$, of the sugar derivative, as confirmed by FAB-MS and SSIMS analysis of the galactose–monomer derivative alone. Thus, SSIMS is able to confirm the surface orientation of the sugar species at the colloid surfaces, albeit in a qualitative manner, and in this case the data again showed an excellent correlation with the quantitative elemental composition by XPS and colloid surface potential measurements. In similar studies, Thompson and coworkers[77] have employed SSIMS to confirm the presence of lysolecithin (lipid) molecules and the nature of their immobilization to a silanized silicon biosensor substrate.

The surface grafting of polymers to a surface assumes a particular significance in polymer colloid chemistry. The adsorption or covalent grafting of hydrophilic polymers such as poly(ethylene oxide) have been employed to stabilize colloids by steric stabilization. The SSIMS and XPS analysis of poly(styrene) colloids copolymerized with a comonomer/stabilizer (macromonomer) methoxypoly-(ethylene glycol or PEG) methacrylate[78] readily detected the presence of the PEG chains at the surface. For this colloid, there is a dramatic increase in the level of the ether environment, $\underline{C}$—O, within the XPS carbon 1s peak attributable to the presence of surface PEG, e.g. chains, compared to polystyrene colloids prepared with charged sulphate end groups. Similarly, the SIMS data show an increase in the relative dominance of the diagnostic ions of PEG, with increasing macromonomer bulk composition (which are not observed within the poly(styrene) charged latex), i.e. $CH_3CH=OH^+$, $HOCH_2C=O^+$ and $CH_3OCH_2C=O^+$ at m/z 45, 59 and 73, respectively.

These findings may have a particular relevance in advanced drug delivery as the modification of the particle surface with such a hydrophilic coating using poly(ethylene oxide) or PEO containing block copolymers provides a method of targeting colloidal particles to specific sites *in vivo*.[9,71–72] On the injection of an uncoated poly(styrene) colloid into the bloodstream, proteins will adsorb rapidly to the particle surface, leading to the rapid uptake by the liver macrophages cells within a matter of minutes. Intriguingly, the protein adsorption is very significantly reduced for the colloids coated with PEO containing

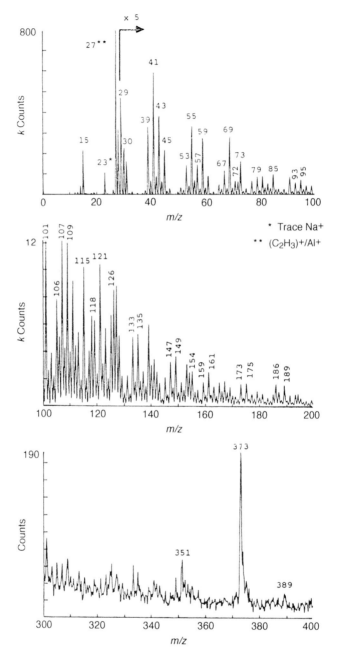

Figure 8. Positive ion ToF-SIMS spectrum (*m/z* 0–400) of a poly(methyl methacrylate/-galactose monomer) copolymer latex prepared with 25% wt/wt galactose derivative[76]

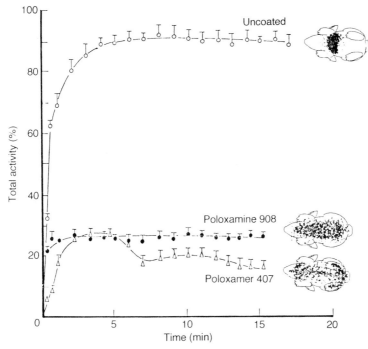

Figure 9. The plot of total activity of ^{125}I in the liver and spleen after administration of ^{125}I labelled 60 nm poly(styrene) colloids that are either uncoated or coated with two commercial PEO-containing surfactants, Poloxmer 908 or 407. The whole body gamma scintigraphy scans are also shown against each plot. The uncoated poly(styrene) colloid is shown localized in the liver. The poloxamine 908-coated system resides within the circulation and is distributed throughout the body. In contrast, the Poloxamer 407-coated colloids are taken up by the bone marrow and outlines the rabbit skeleton (Courtesy of S. S. Davies and L. I. Illum)

surfactant copolymers and the particles are not scavenged by the liver macrophages. Depending on the length of the PEO chain, the particles remain either circulating within the bloodstream or are taken up by the bone marrow or other organs, as shown in Figure 9. Despite such dramatic changes in biodistribution with changes in the PEO copolymer structure, XPS and SSIMS analysis of these surfactant-coated colloids are identical, simply demonstrating significant levels of PEO at the colloid interface.[79]

These findings highlight the problem cited earlier of examining highly hydrated polymer surfaces. In the case of the PEO surfactant-coated poly(styrene) colloids, the change in the block copolymer composition will produce a change in the surface density, conformation and thickness[79] of the PEO steric barrier, which will induce the remarkable changes in protein adsorption and biodistribution. However, when examined in the UHV of the SIMS and XPS instruments, the PEO layer collapses and is indistinguishable for the colloid series. Thus, one must be particularly careful in interpreting data from surface analysis of such systems and be aware of the relevance of hydration in SSIMS and XPS analysis.[80]

4.4 Detection of Organic Species at Polymer Surfaces

Low molecular weight species frequently yield diagnostic secondary ions including $[M \pm H]^{-/+}$ where M corresponds to the molecule's molecular weight. While XPS is able to monitor the presence of molecule within a surface by a change in the atomic percentage or the analysis of an element not present within the substrate, SSIMS frequently provides a molecular fingerprint of organic molecule. However, the major limitation of the SIMS analysis is the qualitative and rarely quantitative nature of the information.[28] In this area, SIMS may be exploited within biomedical science from two primary perspectives: the analysis of bioactive and biological molecules within surfaces and the analysis of polymer surface contaminants.

Analysis of the surface of drug molecules within delivery devices provides information on the surface versus bulk drug composition. Despite the intense interest in the examination of biomaterials by surface analysis techniques, too few reports exist on the analysis of drug delivery systems. SIMS analysis of polymer beads containing 20% wt/wt theophyline show many diagnostic ions observed within the spectra of the pure drug including $(M + H)^+$ and $(M - H)^-$ in the positive and negative (see Figure 10) ion SIMS spectrum, respectively.[81] Benninghoven and his coworkers have pioneered the application of ToF-SIMS for the molecular determination and structural elucidation of low concentrations (in the femtomol range) of a range of large, non-volatile and thermally labile biomolecules with low ion doses.[82–85]

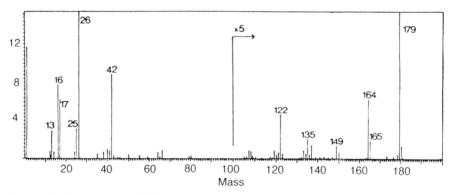

Figure 10. Negative ion SIMS spectrum[81] of theophylline laden (20% wt/wt) polymer beads showing the diagnostic $(M - H)^-$ at m/z 179

Similar studies have been applied to the analysis of inorganic and organic contaminants at biomaterial interfaces. One concerns the presence of a lubricant agent, bisethylene stearmide, on a commercial poly(ether urethane) PEU surface by the detection of diagnostic molecular ions within the SIMS spectra.[26] The cleaning of a polymer latex system of a stabilizing surfactant, sodium dodecyl sulphate, by various cleaning procedures has been followed successfully by monitoring the relative dominance of ions attributable to the substrate polymer and the dodecyl sulphate molecular ion at m/z 265 in the negative ion SIMS spectra of the colloids.[86] One of the most common surface contaminants of polymer surfaces and particularly biomaterials is poly(dimethyl siloxane), or PDMS. An SIMS spectra of a poly(ether sulphone), or PES, contained ions from both the substrate polymer and PDMS, indicating a patchwise coverage despite the fact that XPS detected only a weak Si 2p signal (0.2% atomic ratio).[26]

5 CHEMICAL IMAGING

Images showing elemental composition, chemical state and molecular species by SIMS and (to a lesser extent) XPS imaging have been demonstrated on a range of materials and ideally complement surface structure information available from scanning and transmission electron microscopies. While the potential applications of such techniques to biomedical science are legion, for a variety of reasons SIMS and XPS imaging of polymers and polymeric biomaterials has been restricted to just a few research groups to date.

The application of SIMS imaging and microanalysis to polymers[26,87] has been reviewed recently. SIMS imaging has been limited to a few reports on the analysis of drugs within polymer beads[81] and a drug-laden bead coated with a thin polymeric membrane.[88] The practical resolution of such studies conducted with quadrupole systems was typically of the order of micrometres. The development of ToF-SIMS has allowed SSIMS imaging to high spatial resolution ($<1\ \mu$m) to be realized.[26,89] Submicrometre images of peptides and other large organic molecules on conducting substrates have been achieved by Briggs and Hearn.[90] The SIMS analysis of biological tissues is more well developed,[91-96] offering the prospect of localizing biologically significant trace elements and possibly larger organic molecules. This technique is reviewed elsewhere,[91-96] but is included here to highlight the potential of such studies to the development and performance of biomedical devices. The recently developed XPS imaging may also develop as an important tool for the chemical imaging of biomaterial surfaces. Current commercial instrumentation is capable of imaging at a practical resolution of 0.5–10 μm with a restricted field of view down to 50 μm.[97] Unfortunately, while the application to biological surfaces has been explored,[98] XPS imaging of polymeric surfaces is not well developed and is currently the subject of intense validation by a number of research groups.

6 CONCLUSIONS

This review has highlighted the broad range of potential applications of SIMS in biomedical science. It is evident that the complimentary nature of the SIMS and XPS data will allow a greater insight into the biomaterial surface chemistry. However, significant work still remains to be undertaken to address and resolve many of the issues discussed within the text. It is anticipated that the emergence of surface analytical techniques such as SIMS, XPS and more recently scanning probe technology will contribute through a multitechnique and multidisciplinary approach to a more detailed understanding of the physicochemical properties of biomedical interfaces. With increasing interest being shown internationally in the surface chemistry of biomaterials, it is hoped that in the future these techniques will provide some further insight into this fascinating field of biomaterials research.

REFERENCES

1. Good, R. J., *Surf. Colloid Sci.*, **11**, 1 (1979).
2. Andrade, J. D., Smith, L. M., and Gregonis, D. E., in J. D. Andrade, (ed.), *Surface and Interfacial Aspects of Biomedical Polymers*, Vol. 1, *Surface Chemistry and Physics*, Plenum Press, New York (1985), p. 246.
3. Andrade, J. D. (ed.), *Surface and Interfacial Aspects of Biomedical Polymers*, Vol. 2, *Protein Adsorption*, Plenum Press, New York (1985).
4. Krull, U. J., Thompson, M., and Wong, H. E., in D. Schuetzle and R. Hammerle (eds.) *Fundamentals and Applications of Chemical Sensors*, ACS Symposium Series 309, American Chemical Society, Washington, DC (1986), p. 351.
5. Thompson, M., and Vandenberg, E. T., *Clin. Biochem.*, **19**, 255 (1986).
6. Rembaum, A., and Yen, S. P. S., *J. Macromol. Sci. Chem.*, **A13**, 603 (1979).
7. Fornusek, L., and Vetvicka, V., *CRC Crit. Rev. Ther. Drug Carr. Syst.*, **2**, 137 (1986).
8. Duncan, R., and Kopacek, J., *Adv. Polym. Sci.*, **57**, 51 (1984).
9. *Site-specific drug delivery*, Eds. Tomlinson, E. and Davis, S. S., J. Wiley & Sons, Chichester (1986).
10. Ruckenstein, E., and Gourisankar, S. V., *Biomaterials*, **7**, 403 (1986).
11. Hupfer, B., and Ringsdorf, H., in J. D. Andrade (ed.), *Surface and Interfacial Aspects of Biomedical Polymers*, Vol. 1, *Surface Chemistry and Physics*, Plenum Press, New York (1985), p. 77.
12. Crohn, D., *Prog. Biomed. Engng*, **5**, 43 (1988).
13. Yasuda, H., and Gasicki, M., *Biomaterials*, **3**, 69, 1982.
14. Hub, H. H., Hupfer, B., Koch, H., and Ringsdorf, H., *J. Macromol. Sci. Chem.*, **A15**, 701 (1981).
15. Johnson, D. S., Sanghera, S., Pons, M., and Chapman, D., *Biochem. Biophys. Acta*, **602**, 57 (1980).
16. Durrani, A. A., Hayward, J. A., and Chapman, D., *Biomaterials*, **7**, 121 (1986).
17. Gossen, M. F. A., and Sefton, M. V., *J. Biomed. Mater. Res.*, **17**, 359 (1983).
18. Mori, Y., Nagaoka, H., Takiuchi, T., Kikuchi, T., Noguchi, N., Tanzawa, H., and Noishiki, Y., *Trans. Am. Soc. Artif. Organs*, **28**, 459 (1982).
19. Miyama, H., Fujii, N., Hokari, N., Toi, H., Nagaoka, S., Mori, Y., Noishiki, Y., *J. Bioact. Compat. Polym.*, **2**, 222 (1987).

20. Harper, G., Davies, M. C., Davis, S. S., Illum, L. I., Taylor, D., Irving, M., and Tadros, Th., *Biomaterials*, **12**, 695 (1991).
21. Swalen, J. D., Allara, D. L., Andrade, J. D., Chandross, E. A., Garoff, S., Israelachvilli, J., McCarthy, T. J., Murray, R., Pease, R. F., Rabolt, J. F., Wynne, K. J., and Yu, H., *Langmuir*, **3**, 932 (1987).
22. Briggs, D. (ed.), *Handbook of X-ray and Ultraviolet Photoelectron Spectroscopy*, Heyden and Son, Philadelphia (1977).
23. Ratner, B. D., *Ann. Biomed. Engng*, **11**, 313 (1983).
24. Andrade, J. D., in J. D. Andrade (ed.), *Surface and Interfacial Aspects of Biomedical Polymers*, Vol. 1, *Surface Chemistry and Physics*, Plenum Press, New York (1985), p. 105.
25. Briggs, D., *Surf. Interf. Anal.*, **9**, 391 (1986).
26. Briggs, D., in W. J. Feast and H. S. Munro (eds.), *Polymer Surfaces and Interfaces*, Wiley, Chichester (1987), p. 33.
27. Briggs, D., *Br. Polym. J.*, **21**, 3 (1989).
28. Davies, M. C., and Lynn, R. A. P., *Crit. Rev. Biocompat.*, **5**(4), 297 (1990).
29. Knutson, K., and Lyman, D. J., in J. D. Andrade (ed.), *Surface and Interfacial Aspects of Biomedical Polymers*, Vol. 1, *Surface Chemistry and Physics*, Plenum Press, New York (1985), p. 197.
30. Binnig, G., Rohrer, H., Gerber, Ch., and Weibel, E., *Phys. Rev. Lett.*, **49**, 57 (1982).
31. Vickerman, J. C., Brown, A., and Reed, N. M., *Secondary Ion Mass Spectrometry—Principles and Applications*, Oxford University Press (1989).
32. Werner, H. W., *Surf. Interf. Anal.*, **2**, 56 (1980).
33. Steffens, P., Niehuis, E., Friese, T., Greifendorf, D., and Benninghoven, A., *J. Vac. Sci. Technol.*, **A3**, 1322 (1985).
34. Briggs, D., and Hearn, M. J., *Vacuum*, **36**, 1005 (1986).
35. Hearn, M. J., and Briggs, D., *Surf. Interf. Anal.*, **9**, 411 (1986).
36. Hearn, M. J., Briggs, D., Yoon, S. C., and Ratner, B. D., *Surf. Interf. Anal.*, **10**, 384 (1987).
37. Briggs, D., *V. G. Sci. Newsletter*, **2** (Spring 1986).
38. Brown, A., and Vickerman, J. C., *Surf. Interf. Anal.*, **8**, 75 (1986).
39. McLafferty, F. W., *Interpretation of Mass Spectra*, 2nd edn, W. A. Benjamin Inc., Reading, Ma. (1974).
40. Short, R. D., and Davies, M. C., *Int. J. Mass Spectrosc. Ion Proc.*, **89**, 149 (1989).
41. Briggs, D., *Org. Mass Spectrom.*, **22**, 91 (1987).
42. Davies, M. C., Short, R. D., Khan, M. A., Watts, J. F., Brown, A., Eccles, A. J., Humphrey, P., Vickerman, J. C., and Vert, M., *Surf. Interf. Anal.*, **14**, 115 (1989).
43. Hearn, M. J., and Briggs, D., *Surf. Interf. Anal.*, **11**, 198 (1988).
44. Brinkhuis, R. J., and van Ooij, W. J., *Surf. Interf. Anal.*, **11**, 214 (1988).
45. Briggs, D., and Munro, H. S., *Polym. Commun.*, **28**, 307 (1987).
46. Briggs, D., Chilkoti, A., and Ratner, B. D., Preliminary results.
47. Chilkoti, A., Ratner, B. D., and Briggs, D., *Chem. Mater.*, **3**, 51 (1991).
48. Leggett, G. J., Briggs, D., and Vickerman, J. C., *J. Chem. Soc. Faraday Trans.*, **86**, 1863 (1990).
49. Leggett, G. J., Vickerman, J. C., and Briggs, D., *Surf. Interf. Anal.*, **16**, 3 (1990).
50. Leggett, G. J., Briggs, D., and Vickerman, J. C., *Surf. Interf. Anal.*, **17**, 737 (1991).
51. Davies, M. C., Khan, M. A., Short, R. D., Aktar, S., and Pouton, C., *Biomaterials*, **11**, 228 (1990).
52. Beamson, G., Bunn, A., and Briggs, D., *Surf. Interf. Anal.*, **17**, 105 (1991).
53. Beamson, G., and Briggs, D., *Surf. Interf. Anal.*, **15**, 541 (1990).
54. Davies, M. C., Lynn, R. A. P., Heller, J., and Paul, A., *Macromolecules*, **24**, 5508 (1991).

55. Hearn, M. J., Ratner, B. D., and Briggs, D., *Macromolecules*, **21**, 2950 (1988).
56. Briggs, D., and Ratner, B. D., *Polym. Commun.*, **29**, 6 (1988).
57. Briggs, D., Hearn, M. J., and Ratner, B. D., *Surf. Interf. Anal.*, **6**, 184 (1984).
58. Lub, J., van Vroonhaven, F. C. B. M., van Leyen, D., and Benninghoven, A., *J. Polym. Sci., Polym. Chem. Edn*, in press.
59. Davies, M. C., Khan, M. A., Brown, A., and Humphrey, P., in A. Benninghoven, A. M. Huber and H. W. Werner (eds.), *SIMS VI*, John Wiley, Chichester (1988), p. 667.
60. Blestos, I. V., Hercules, D. M., Greifendorf, D., and Benninghoven, A., *Anal. Chem.*, **57**, 2384 (1985).
61. Hercules, D. M., and Blestos, I. V., *Polym. Mater. Sci. Eng*, **54**, 302 (1986).
62. Hercules, D. M., in A. Benninghoven, A. M. Huber and H. W. Werner (eds.), *SIMS VI*, John Wiley, Chichester (1988), p. 599.
63. Fakes, D. W., Davies, M. C., Brown, A., and Newton, J. M., *Surf. Interf. Anal.*, **13**, 233 (1988).
64. Fakes, D. W., Newton, J. M., Watts, J. F., and Edgell, M. J., *Surf. Interf. Anal.*, **10**, 416 (1987).
65. van Ooij, W. J., Brinkhuis, R. H. G., and Newman, J., in A. Benninghoven, A. M. Huber and H. W. Werner (eds.), *SIMS VI*, John Wiley, Chichester (1988), p. 671.
66. Lub, J., van Vroonhoven, F. C. B. M., van Leyen, D., and Benninghoven, A., *Polymer*, **29**, 998 (1988).
67. Lub, J., van Vroonhoven, F. C. B. M., Bruninx, E., and Benninghoven, A., *Polymer*, **30**, 40 (1989).
68. Ochiello, E., Morra, M., Garbassi, F., and Bagon, J., *Appl. Surf. Sci.*, **36**, 285 (1989).
69. Chilkoti, A., and Ratner, B. D., *Surf. Interf. Anal.*, **17**, 567 (1991).
70. Brinen, J. S., Greenhouse, S., and Jarrett, P. K., *Surf. Interf. Anal.*, **17**, 259 (1991).
71. Illum, L., Jacobsen, L. O., Muller, R. H., Mak, E., and Davis, S. S., *Biomaterials*, **8**, 113 (1987).
72. Davis, S. S., Douglas, S. J., Illum, L., Jones, P. D. E., Mak, E., and Muller, R. H., in G. Gregoriadis and J. Senior (eds.), *Targeting of Drugs with Synthetic Systems*, Plenum, New York (1986), p. 123.
73. Lynn, R. A. P., Davis, S. S., Short, R. D., Davies, M. C., Vickerman, J. C., Humphrey, P., Johnson, D., and Hearn, J., *Polym. Commun.*, **29**, 365 (1988).
74. Davies, M. C., Lynn, R. A. P., Davis, S. S., Hearn, J., Watts, J. F., Vickerman, J. C., and Johnson, D., *J. Colloid Interf. Sci.*, in press.
75. Davis, S. S., Illum, L., McVie, J. G., and Tomlinson, E. (eds.), *Microspheres and Drug Therapy: Pharmaceutical, Immunological and Medical Aspects*, Elsevier, Amsterdam (1984).
76. Davies, M. C., Lynn, R. A. P., Davis, S. S., Hearn, J., Vickerman, J. C. and Paul, A. J., *J. Colloid Interf. Sci.*, submitted for publication.
77. Ghaemmaghami, V., Kallury, K. M. R., Krull, U. J., Thompson, M., and Davies, M. C., *Anal. Chem. Acta*, **225**, 369 (1989).
78. Brindley, A., Davies, M. C., Lynn, R. A. P., and Watts, J. F., *Polym. Commun.*, **33**, 1112 (1992).
79. Muir, I., Moghimi, M., Illum, L. I., Davis, S. S., and Davies, M. C., submitted to *Pharm. Res.*
80. Ratner, B. D., in B. D. Ratner (ed.), *Surface Characterisation of Biomaterials*, Progress in Biomedical Engineering Series 6, Elsevier, New York (1988), p. 15.
81. Davies, M. C., Brown, A., Newton, J. M., and Chapman, S. R., *Surf. Interf. Anal.*, **11**, 591 (1989).
82. Lange, W., Griefendorf, D., van Leyen, D., Niehuis, E., and Benninghoven, A., in A. Benninghoven, R. J. Colton, D. S. Simons and H. W. Werner (eds.), *SIMS V*, Springer Series in Chemical Physics 44 (1986), p. 67.

83. Benninghoven, A., Niehuis, E., Friese, T., Griefendorf, D., and Steffens, P., *Org. Mass Spectrom.*, **19**, 346 (1984).
84. van Leyen, D., Greifendorf, D., and Benninghoven, A., in A. Benninghoven, A. M. Huber and H. W. Werner (eds.), *SIMS VI*, John Wiley, Chichester (1988), p. 679.
85. Benninghoven, A., and Sichtermann, W. K., *Anal. Chem.*, **50**, 1180 (1978).
86. Koosha, F., Muller, R. H., Davis, S. S., and Davies, M. C., *J. Control. Rel.*, **9**, 149 (1989).
87. Brown, A., and Vickerman, J. C., *Analyst*, **109**, 851 (1984).
88. Davies, M. C., and Brown, A., in P. I. Lee and W. R. Good (eds.), *Controlled Release Technology, Pharmaceutical Applications*, ACS Symposium Series 348, Washington, DC (1987), p. 100.
89. Briggs, D., Hearn, M. J., Beamson, G., and Fletcher, I., *Spectrosc. World*, **2**(6), 11 (1990).
90. Briggs, D., and Hearn, M. J., *Surf. Interf. Anal.*, **13**, 181 (1988).
91. Linton, R., in A. Benninghoven, R. J. Colton, D. S. Simons and H. W. Werner (eds.), *SIMS V*, Springer Series in Chemical Physics 44 (1986), p. 420.
92. Burns, M. S., in A. D. Romig and W. F. Chambers (eds.), *Microbeam Analysis*, San Francisco Press (1986), p. 193.
93. Galle, P., and Berry, J. P., in R. Feder, J. W. McGowan and D. M. Shinozaki (eds.), *Examining the Submicron World*, Plenum, New York (1986), p. 35.
94. Burns, M. S., File, D. M., Deline, V., and Galle, P., *Scanning Electr. Micr.*, **4**, 1277 (1986).
95. Burns, M. S., in A. D. Romig and W. F. Chambers (eds.), *Microbeam Analysis*, San Francisco Press (1986), p. 95.
96. Galle, P., *Ann. Phys. Fr.*, **10**, 287 (1985).
97. Briggs, D., Hearn, M. J., Beamson, G., and Fletcher, I., *Spectrosc. World*, **2**(6), 11–15 (1990).
98. Griffith, O. H., *Appl. Surf. Sci.*, **26**, 265–279 (1986).
99. Davies, M. C., Langer, R., Khan, M. A., Watts, J., and Domb, A., *J. Appl. Polym. Sci.*, **42**, 1597 (1991).
100. Davies, M. C., Short, R. S. Vickerman, J. C., Humphrey, P., and Brown, A., *Polym. Mater. Sci. Eng*, **59**, 729 (1988).
101. Lub, J., van Leyen, D., and Benninghoven, A., *Polym. Commun.*, **30**, 74 (1989).
102. Briggs, D., *Surf. Interf. Anal.*, **4**, 151 (1982).
103. van Ooij, W. J., and Brinkhuis, R. H. G., *Surf. Interf. Anal.*, **11**, 430 (1988).
104. Briggs, D., *Surf. Interf. Anal.*, **8**, 133 (1986).
105. Lub, J., and Benninghoven, A., *Org. Mass Spectrom.*, **24**, 164 (1989).
106. Lub, J., van Velzen, P. N. T., van Leyen, D., Hagenhoff, B., and Benninghoven, A., *Surf. Interf. Anal.*, **12**, 53 (1988).
107. Briggs, D., *Surf. Interf. Anal.*, **5**, 113 (1987).

9

Scanning Probe Microscopy—Current Issues in the Analysis of Polymeric Biomaterials

M. C. Davies, D. E. Jackson, C. J. Roberts,
S. J. B. Tendler, K. M. Kreusel, M. J. Wilkins and
P. M. Williams

Department of Pharmaceutical Sciences,
University of Nottingham

1 SCANNING TUNNELLING MICROSCOPY—BACKGROUND

Since its invention by Binnig and Rohrer[1] in 1982, the scanning tunnelling microscope (STM will henceforth denote scanning tunnelling microscope or microscopy) has become a powerful tool for the study of conducting surfaces to atomic resolution.[2-7] While initially STMs operated in an ultra-high vacuum environment (UHV), it soon became clear that imaging under ambient conditions[8] or in liquids is also possible.[9] This ability to work in non-UHV or aqueous environments encouraged the development of the STM for imaging biological molecules. In this capacity, it is envisaged that the STM may complement the present range of biophysical techniques, such as electron microscopy, computational chemistry, high field n.m.r. spectroscopy and X-ray crystallography which are employed to investigate biomolecular structure.

While there is a concerted effort in the application STM to biological specimens, the technique is very much in its infancy. Many critical issues and analysis procedures remain to be resolved. A number of comprehensive reviews of the STM analysis of biomolecules has been published recently.[8,10,11] In this chapter, we shall briefly describe the principle of the technique, and concisely highlight the current work in the biomedical field. We will concentrate on the problems facing STM in its attempt to establish itself as a reproducible and reliable method, and assess the success of recent work to overcome some of these obstacles.

Polymer Surfaces and Interfaces II
Edited by W. J. Feast, H. S. Munro and R. W. Richards
© 1993 John Wiley & Sons Ltd

2 STM—THEORY AND INSTRUMENTATION

The STM's exceptional resolution is achieved by monitoring the electric current which flows, on the application of a voltage bias, across a very small gap between an atomically sharp probe and a substrate. Typical values for the imaging parameters are 10 pA–1 nA for the current (J), 1 mV–1 V for the voltage (V) and approximately 0.3–1 nm for the gap between the tip and the sample (s). The transition of the electrons between the two electrodes is due to a quantum mechanical tunnelling effect, which may be modelled approximately for a vacuum tunnelling junction due to Simmons:[12]

$$J \alpha \frac{V}{s} \exp(-s\phi^{1/2}) \tag{1}$$

where ϕ represents the average of the work functions of the probe and the sample. A more rigorous analysis of STM imaging theory may be found in the literature.[13–15]

The very high vertical resolution achievable with an STM (better than 10 pm) results from the exponential dependence of J upon s, as indicated in equation (1), which gives approximately an order of magnitude change in the current for a 0.1 nm alteration in the electrode separation.[16] Resolution in the horizontal plane (better than 100 pm) results from the confinement of the current filament to the very apex of the probe, which ideally consists of a single atom.

To build up a three-dimensional 'topographic' surface map the probe is scanned in a raster-like manner above the surface while maintaining the tunnelling current at a constant pre-set value (constant current mode, see Figure 1(a)) by means of a feedback loop. The resultant control signal (shown in Figure 1(b)) contains data which are derived from a convolution of the tip and surface profile.[16] However, due to the fact that the tip is normally considerably sharper than any surface features, the sample topography dominates the image.

The control of the spatial position of the tip within the subangström limits required is achieved by making use of scanning elements constructed from piezo-ceramic crystal. This material can be made to controllably expand and contract at a subatomic level on the application of appropriate voltage biases (± 500 V). The most common form of scanning head consists of a piezo-ceramic tube, with the probe mounted axially at one end (shown schematically in Figure 1(a)). By supplying appropriate voltages to quartered metallic electrodes on the tube's outer surface and to a single electrode on the inner surface, the scanner flexes and expands to give the necessary control. The development of this system has now reached such a stage that, despite its serial nature, the 'real' time imaging with an STM has been achieved.[17]

More complete discussions of STM design, including methods of overcoming vibration, piezo-ceramic hysteresis and thermal drift problems, have been

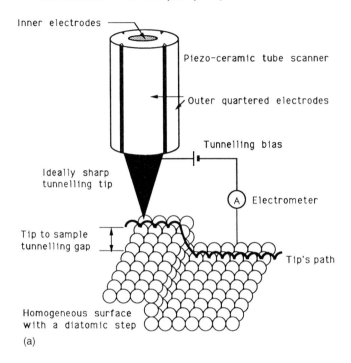

Inner electrodes

Piezo-ceramic tube scanner

Outer quartered electrodes

Tunnelling bias

Ideally sharp
tunnelling tip

(A) Electrometer

Tip to sample
tunnelling gap

Tip's path

Homogeneous surface
with a diatomic step

(a)

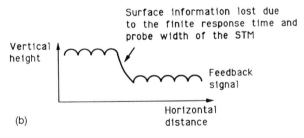

Surface information lost due
to the finite response time and
probe width of the STM

Vertical
height

Feedback
signal

Horizontal
distance

(b)

Figure 1. (a) A schematic diagram showing (not to scale) the STM scanning head, consisting of a piezo-ceramic tube and an atomically sharp tip, imaging a homogeneous surface containing a diatomic step in the 'constant current' mode. (b) A representation of the feedback signal that would be recorded from the scan line shown in (a). Notice the loss of information at the step edge, due to the time delay in the response of the feedback loop and a convolution of the tip profile into the data

reported.[18–20] A number of inexpensive STM instruments are commercially available.[21–23]

On a practical level, biological samples are usually prepared for STM analysis by deposition on to a suitable conducting substrate from solution and allowing to dry either by evaporation in air or by freeze drying. To date, highly oriented pyrolytic graphite (HOPG) and gold have been the most commonly employed substrates, due to their being nearly atomically flat over micrometre-sized regions and resistant to contamination in air. Due to the insulating properties of large biological molecules, the use of sputtered or evaporated metallic coatings has also been advocated to ensure surface conductance.

3 STM OF POLYMERIC BIOMOLECULES AND BIOMATERIALS

Since the first images of DNA[24] and a bacteriophage head[25] were published in 1984 and 1985, respectively, there have been many papers devoted to the STM imaging of a wide range of biological molecules.[26–41] Figure 2 shows a 350×350 nm STM image of a HOPG surface on to which purified ovarian carcinoma polymorphic epithelial mucin (PEM) has been deposited.[42] The PEM molecules are high molecular weight glycoprotein biopolymers that associate to provide protection to the epithelial cell surface and may afford the malignant cell a selective advantage for growth. STM analysis reveals that the glycoprotein molecules aggregate into sheet-like structures, and in Figure 2 these aggregates appear to be composed of associated rod-shaped molecules whose length and width range from 25 to 45 nm and 3 to 4 nm, respectively.

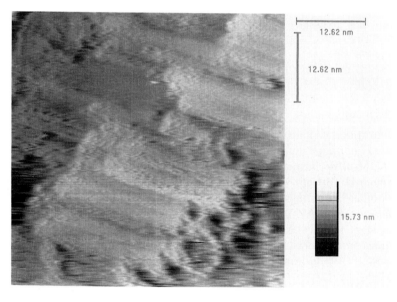

Figure 2. STM micrographs (50 nm × 50 nm) of purified human ovarian carcinoma PEM. The molecules can be seen as polymorphic rods on the HOPG substrate

The polymorphic nature of the PEM sample is reflected in the range of the molecular dimensions observed within the STM images; these data correlate well with previous electron microscopy studies. Similar reports on molecular resolution images have been reported for a range of biomolecules including proteins, carbohydrates, poly(nucleic acids) and polymers. To date, atomic resolution images of biomolecules and organic species have been limited to images of DNA[43] and liquid crystals.[45,46] A brief summary of the more notable publications in STM analysis of biomolecules and biopolymers is provided in Table 1.

In an attempt to validate and interpret STM data, there is an increasing interest being shown in the comparison of the molecular conformation observed within the STM image to known structural parameters derived from conventional biophysical techniques such as X-ray crystallography and n.m.r. spectroscopy. To achieve this aim, image manipulation procedures and computational techniques need to be exploited and, as such, the integration of computational chemistry, molecular modelling and STM image analysis is a valuable approach. A particularly notable use of molecular graphics in the interpretation of molecular conformation in STM images has been reported by Smith and his coworkers in an atomic resolution study of the liquid crystalline 4-n-alkyl-4-cyanobiphenyl molecules.[45] By constructing a CPK molecular model with a lattice of holes with a spacing of 2.456 Å to represent the graphite surface, the authors have shown the two-dimensional organization of the molecules observed within the STM image to be due to the registry of the alkyl chains with the substrate atoms. There was excellent agreement between the unit cell width between the model and the STM data, e.g. 38.5 and 38 Å, respectively, for the 4-n-octyl-4-cyanobiphenyl. To achieve such a fit, it was necessary to tilt the aromatic rings in the model, which was also confirmed by a contour map profile over the phenyl groups in the STM image. The authors have also noted the higher contrast shown for the aromatic groups than for the alkyl chains in the STM image. In *ab initio* Hartree–Fock molecular orbital calculations of the highest occupied (HOMO) and lowest unoccupied (LUMO) molecular orbitals, the aromatic carbons and nitrogen were found to have the highest HOMO and LUMO densities. Therefore, the molecular orbital calculations suggested that the distinction between the alkyl chains and the cyanobiphenyl moieties in the STM images is directly related to a number of differences in molecular orbital densities, i.e., the contrast mechanism is not purely topographical but may be influenced by electron densities.

Despite the considerable progress in the applications of STM in biomedicine and polymer science, there are a number of problems to be resolved before the technique may be routinely applied at even molecular resolution to biomolecular analysis. From a fundamental perspective, it must be noted that the method of image formation is at present little understood. The means by which an appreciably large current is made to flow 'through' an insulating biomolecule

Table 1. A small selection of STM results of biomolecules published to date

Type molecule	Environment of experiment	Sample preparation	References
Proteins/ peptides	Air	PBLG on HOPG	26
		Collagen on HOPG	27
		Phosphorylase b and kinase	28
		Mucin on HOPG	42
Carbohydrates	Air	Cyclodextrin on HOPG	29
		Cellulose and xanthan on HOPG	30
DNA	Air	Air dried on HOPG	31–33
		Freeze dried on Pt/C and coated with Pt/C/Ir	34
		Freeze dried on $MgCl_2$	35
	UHV	Air dried on HOPG	43
	Water	in solution on Au	36
	Oil	Air dried on HOPG	31
	Salt solution	Electrochemically deposited on HOPG and Au	37
Polymers	Air	PCPS and PDA air dried on HOPG and Au	38
		LB layers of PODA and PMMA on HOPG	39
		Polyethylene on mica, coated in Pt/C	40
Viruses	Air	Dried on Au and coated in Pt/C	44
Membranes/ phospholipids	Air	LOPC on chemically modified HOPG	47
		Porin on amorphous carbon	48

poses severe theoretical difficulties.[49–52] Perhaps the most significant problem arises from the fact that many STM research groups are finding that there is frequently a lack of reproducibility in the STM imaging of biomolecules and polymers.[11] The disconcerting range of molecular dimensions quoted for published STM studies on DNA bears testament to this problem. There are many factors which may contribute to this lack of reproducibility, including the different sample preparation and tip fabrication procedures and the analysis conditions employed. However, there are two specific practical problems that undermine the routine imaging of biomolecules: tip-induced movement of the biomolecule and the detection of artefacts.

3.1 Biomolecular Movement

For stable STM images to be obtained, the biomolecule needs to adhere to the conducting substrate, and this adherence will depend on a number of factors, including their relative surface energies and hydrophobicities.[53] As shown schematically in Figure 3(a), where the molecule–substrate adhesion is too low, the STM tip may simply 'sweep' it over the substrate and no stable image will be obtained.[42,54,55] This effect has been reported in the STM imaging of biomolecules and gold islands.[56] Similarly, the molecule may be picked up by the tip itself during the scanning process. The problem is compounded by the fact that biomolecules are usually considerably larger than the normal tunnelling gap of 5–10 Å. In the case of imaging in a liquid where there will be significant molecular mobility, the equilibrium adsorption of the molecules at the substrate interface is even more critical. This can be achieved for charge-carrying molecules by operating in an electrolytical cell where the surface potential may be controlled.[57]

The problem of improving biomolecule adsorption to and stability at the substrate surface has been approached in several ways.[34,45,46,58–64] The main aims of these techniques are to produce a suitable surface coverage of the molecule, to avoid aggregation if possible and to stabilize the molecule so that it may be imaged.

To date, there have been relatively few types of substrates employed in STM experiments. Preference has been given to those which may be readily prepared and possess atomically flat regions of square micrometre dimensions. In this respect, both hydrophilic (freshly prepared gold) and hydrophobic (HOPG) surfaces are available and, therefore, one should select the substrate that is both suitable for the STM and to which the molecule of interest shows significant adsorption.[11] An example of this approach is the use of gold for certain proteins such as antibodies.[57] Similarly, STM images of alkyl cyanobiphenyl molecules have been shown to form a two-dimensional lattice on HOPG owing to the registry of the alkyl chains with the graphite.[45] In this way, the molecules are stabilized within the lattice structure and thus do not move when imaged. The

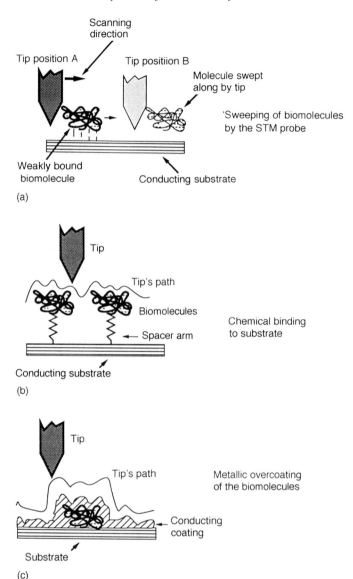

Figure 3. (a) This simplified diagram illustrates one of the major obstacles to obtaining stable images of lightly adsorbed large biomolecules. Since the tunnelling gap is small in comparison with the molecular dimensions, the tip simply sweeps it along the substrate before an image can be obtained. (b) One method of overcoming the problem illustrated in (a) is to chemically bond the biomolecule to the substrate so that it remains fixed as the tip passes over the molecule. (c) An alternative method involves coating molecules with a conducting metallic coat. Although this reduces the potential resolution, it circumvents the conductivity problem inherent in imaging insulating biomaterials

potential influence of the substrate on the spatial arrangement and conformation of adsorbed molecules is illustrated by the change in lattice structure when an alternative substrate, molybdinum disulphide (MoS_2), was employed for the STM imaging of the alkyl cyanobiphenyl molecules.[46]

Chemical modification of the substrate or the sample has been employed in STM analysis to assist in the adsorption or immobilization of molecules to the surface, as illustrated in Figure 3(b). The surfactant, benzyldimethylalkyl ammonium chloride (BAC) has been employed in STM imaging of DNA to improve its surface coverage (i.e. spreading) on gold,[58] although there must be some caution in the use of this approach as BAC has been shown to increase the DNA molecular dimensions. Tris(1-aziridinyl)phosphine oxide (TAPO) has been employed to cross-link DNA and fix it to a gold substrate.[59] Heckl and his coworkers have used biosensor technology to chemically modify an HOPG substrate with silane coupling agents for the immobilization of lysolecithin (lipid) molecules[47] where their outstanding STM images clearly distinguish the individual lipid molecules within the surface grafted monolayer. Another approach makes use of the STM's ability to locally modify a substrate.[60–62] This process has been used to place DNA on to HOPG, by first coating the tip in a solution containing the DNA and then by applying a voltage pulse to the tunnelling junction to locally deposit the molecules.[63] Perhaps the most traditional alternative to these methods is that of covering the molecule on the substrate with a thin conducting layer, usually Pt/C[34] or Pt/C/Ir[34], as shown in Figure 3(c). This has the added benefit of stabilizing the molecules during imaging. This method does, however, reduce the resolution potentially obtainable due to a minimum coating grain size of the order of 2–5 nm. Nevertheless, useful data may be obtained as illustrated by the 1200 nm × 1200 nm topograph in Figure 4. This shows a supramolecular network of xanthan molecules after being coated in Pt/C.[64] Corresponding experiments with uncoated xanthan deposited on graphite were unsuccessful due to a low molecule–substrate interaction and the subsequent tip-induced movement. Xanthan is a large polysaccharide (molecular weight approximately 2×10^6) commonly used in the food, pharmaceutical and colloid industries. The xanthan molecules form an entangled polymer network in solution and it is this entanglement that is suggested by the complex structure observed in the image in Figure 4.

3.2 Artefacts

As in the early history of electron microscopy, a problem that has emerged very recently is the identification of artefacts which can closely mimic expected biomolecular structures, especially if HOPG is used as the substrate,[65] and which further complicate image interpretation. Artefacts on freshly cleaved HOPG in Figure 5(a) and (b) illustrate this point. The occurrence of a helical structure (Figure 5(a)) that resembles DNA and a superlattice-like order similar

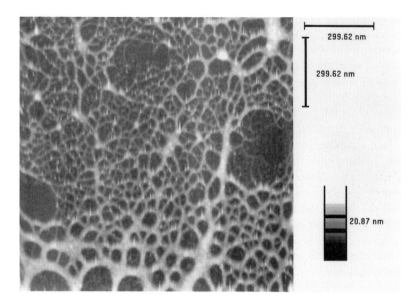

Figure 4. A 1200 nm × 1200 nm image of a xanthan molecular network. This image could only be obtained after coating the molecules, with platinum–carbon in this case, to avoid the tip sweeping the molecular network across the HOPG substrate

to that proposed for protein crystals (Figure 5(b)) highlights the significant care that must be taken in interpreting biomolecular images on an HOPG substrate.

As yet, the methodologies aimed at distinguishing artefacts in STM images are not well established. While making extensive studies of bare substrates to catalogue anomalous features is useful and necessary, there will always be some doubt as to the nature of a feature thought to be an adsorbed molecule in ordinary 'topographic' images. Therefore, a number of approaches have been advocated which may shed some light on this issue.

The correlation of the STM images with conformational data obtainable from other biophysical techniques such as X-ray crystallography, high-field n.m.r. spectroscopy and computational chemistry has already been noted above as an important goal for the technique's development for polymer and bio-molecular analysis.[66] Where major discrepancies exist between the data sets, there must be some concern as to the nature of the feature. However, some care must be exercised in such correlations with other data sets since an STM would not be expected to produce an image of a molecule resulting from its surface topography, but from its electron density distribution when bound to a substrate.[67] Similarly, the deviations from X-ray crystallography data may also arise in STM analysis due to the influence of a variety of factors including tip geometry, the existence of surfactants, buffers, etc.

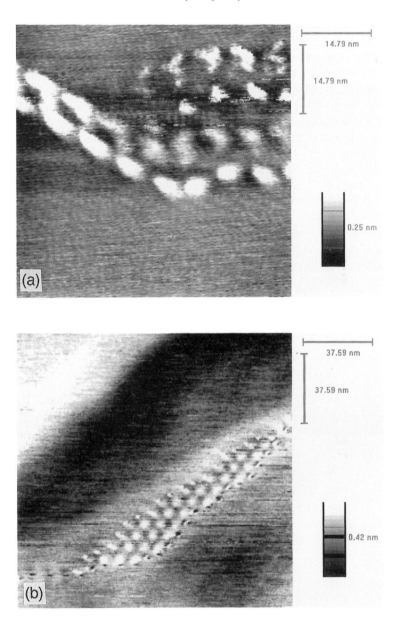

Figure 5. Two types of artefacts observed on freshly cleaved naked HOPG which can confuse image interpretation. (a) This is a helical artefact; the scan area is 59 nm × 59 nm and was recorded at a gap resistance of 310 MΩ. The width of the apparent chain is 3.6 nm and its periodicity 1.9 nm. Secondly, an apparent superlattice network may be observed as in (b), which is a 150 nm × 150 nm scan

Binnig and coworkers have noted that for HOPG the direction of the graphite lattice around linear artefacts may give a clue to the origin of the image.[68] Others have suggested that the correlation of the STM images with those obtained with other imaging techniques such as scanning electron microscopy (SEM) may aid image interpretation.[69] As such artefacts have yet to be reported in STM studies on HOPG coated with a conducting layer such as Pt/C or Pt/C/Ir, the detection of the same features on the native and overcoated substrates (albeit at lower resolution) suggests that they are sample related and not artefactual. STM is also capable of recording several data formats concurrently with the standard 'topographic' image, including spectroscopic,[16] barrier height,[16,59] and tunnelling current fluctuation[66] information, and further studies may shed some light on how these approaches may also assist in elucidating artefacts.

Recently, more direct techniques based on the ability of the probe to interact with the molecules have been proposed (see above). Hence, if a supposed molecule can be swept or desorbed by the tip, this would suggest that the image topography relates to an adsorbed molecule. Conversely, if the target cannot

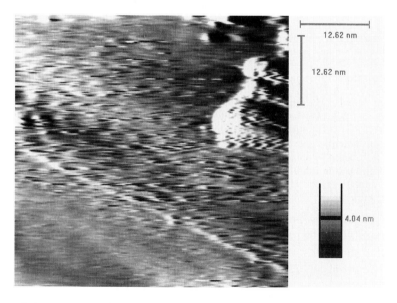

Figure 6. A repeat scan of the same area imaged in Figure 2 (50 nm × 50 nm) using the same imaging parameters. However, between these two scans another had been recorded with the tip closer to the surface. This process has swept most of the mucin molecules from the scan area, thus confirming that the original image was not of an artefactual origin

be removed from an image, there may be some concern that the feature is a substrate-related artefact. An example of this approach may be shown in the confirmation that the features shown in Figure 2 are not surface-related artefacts. Figure 6 is an image[42] of the same area scanned in Figure 2 after the tip has been brought closer to the surface and scanned over the same area. As a result of this, only the underlying HOPG substrate is observed, the mucin molecules having been mostly swept from the scan area by the tip.

An alternative method may be to use the probe to etch a line across the feature. The controlled etching of surfaces has been demonstrated to produce troughs as small as 2 nm wide.[70] An example of two such features is shown in Figure 7. To etch a continuous line in an HOPG surface, the tip is moved from a known starting position to a predetermined location at a set velocity, with the voltage bias raised to, or above, the fabrication level of 3.5 V.[70] Upon reaching the end point of an etched line the tip voltage is then immediately reduced to a non-destructive level. The features produced are then stable in time, unless defoliation of the upper graphite layers is caused by too much scanning of an etched area.[71] Rabe *et al.* have suggested that the presence of water is a necessary condition for the fabrication process to occur,[72] which is

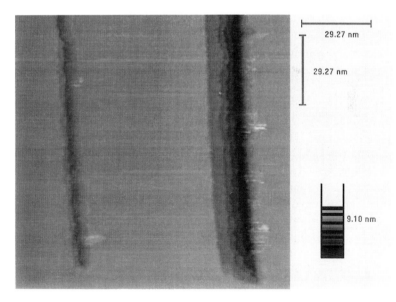

Figure 7. Two trenches produced side by side in HOPG (120 nm × 120 nm). These clearly show the layered nature of the damage. The larger feature was produced with the tip etching parameters set at −4.2 V and 14 nm/s and the thinner one at −3.6 V and 14 nm/s

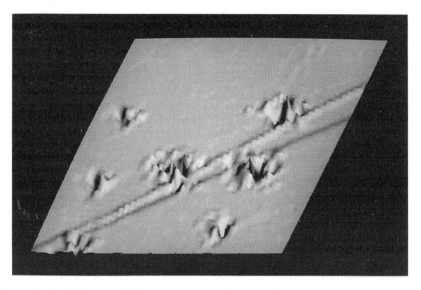

Figure 8. A 708.8 nm × 708.8 nm scan showing a helical artefact which has been identified as such by etching pits at a tip bias of 4.2 V in the HOPG, along its path. Each hole is approximately 100 nm in diameter. If the feature were a biomolecule this process would cause severe disruption of its conformation and probable desorption of the molecule

thought to be based on the following chemical reactions:[71]

$$C_{(solid)} + H_2O_{(liquid)} + 175.32 \text{ kJ/mol} = CO_{(gas)} + H_{2\,(gas)}$$

$$C_{(solid)} + 2\,H_2O_{(liquid)} + 178.17 \text{ kJ/mol} = CO_{2(gas)} + 2\,H_{2\,(gas)}$$

It is the increase in energy density at the surface due to the raised bias voltage that initiates the oxidation of the graphite.

A possible application of this process is shown in Figure 8, where an artefact on bare HOPG, which may be mistaken for a helical biomolecule such as DNA, has been cut at three points by etching a network of pits into the graphite.[66,70] Since the feature remains undisturbed along the axis of the helical structure, this strongly suggests that it arises from a surface artefact. A similar process may also be carried out on Au[54] and Pt/C.

4   OTHER SCANNING PROBE MICROSCOPY (SPM) TECHNIQUES

The immediate success of STM has led to a proliferation of derivative techniques based upon the principle of locally sampling an exponentially decaying field with a probe spatially controlled by a piezo-ceramic scanner. The first and most important of these is the atomic force microscope (AFM).[73] Here an atomically sharp tip, often made of diamond or silicon nitride, is drawn across a surface and kept in contact by a very 'soft' spring (contact mode) (see Figure 9). The tip is then raster scanned in a similar fashion as to the STM, while maintaining a constant force between the probe and the sample. The deflections of the spring, or cantilever, are monitored either using an STM[73] or more commonly by interferometry methods.[74] The AFM thus records contours of constant force due to the repulsion generated by an overlap between the electron clouds of the tip and the surface atoms. One advantage the AFM has over the STM is that the sample may be an insulator, and hence the imaging of biological material, even in its living state,[75] should be less problematic.[76]

Like the STM, an AFM is capable of atomically resolving a surface,[77,78] however, it also suffers from the problem of sweeping molecules across a substrate due to the considerable interaction force that can occur between the tip and surface.[79] Despite this, many isolated organic structures have been resolved to nanometre resolution, including red blood cells and antibodies on a glass substrate[80] and amino acid crystals.[81] In an attempt to reduce the interaction force at the tip-to-surface contact region, most modern biological

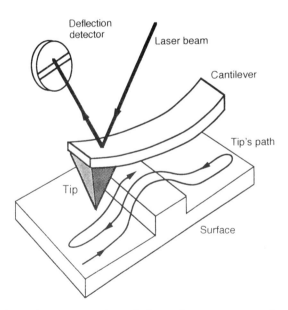

Figure 9. A schematic representation of the main components of an atomic force microscope. A very sharp probe is brought into light contact to a surface by a very soft spring or cantilever. The tip is then tracked across the surface as shown, while keeping the contact force constant, and the defections of the cantilever recorded using a reflected laser beam. This signal then reveals topographic detail in the subnanometre range

AFMs operate in aqueous environment;[82] this negates any capillary forces resulting from the thin film of surface water that covers samples exposed to air.[82] This type of environment has also been used to observe the polymerization of fibrinogen taking place upon a mica surface.[82] Again as with STM the AFM's ability to interact with the surface under study has recently been put to good use to first image a hepatic gap junction and then to dissect it to reveal the internal hemichannels.[83] The recent instrumental advance, which allows the recording of tunnelling current as well as surface forces with the same tip,[84] will ease the task of interpreting biological images and distinguishing artefactual features by supplying both types of data sets.

Other related instruments to the STM and AFM include the scanning near-field optical microscope (SNOM),[85] the scanning thermal microscope,[86] and the scanning chemical potential microscope.[87] As yet these newer microscopes have yet to be applied to the biological field, due to their less well-developed technologies. It is likely, however, that the SNOM and the others will soon follow the success of the STM and AFM.

5 CONCLUDING REMARKS

The results discussed here clearly demonstrate the fact that STM and AFM have much to offer in the biological field. The recent advent of commercial instruments which combine scanning probe techniques with more traditional analysis technology points the way to the future, where STM data will complement that provided by electron microscopy, XPS, SIMS and other surface analysis methods. In addition, the developments briefly illustrated here in computational chemistry which can simultaneously display many different data types will become an important part of the analysis process.

This chapter has provided a snapshot of some of the major issues which at present preoccupy the STM/AFM community in its effort to apply scanning probe technology to biological problems. Some of the work described requires further investigation to realize the true capabilities and limitations of the techniques. In reviewing some of the remarkable results obtained to date we hope that we have highlighted the reason for the considerable effort that is at present underway to apply SPM to biological problems.

ACKNOWLEDGEMENTS

We would like to acknowledge the support of the SERC Protein Engineering/ DTI Link programme, the MAFF/DTI Link programme, Glaxo Group Research Ltd. and VG Microtech.

REFERENCES

1. Binnig, G., Rohrer, H., Gerber, Ch., and Weibel, E., *Phys. Rev. Lett.*, **49** 57–61 (1982).
2. Hallmark, V. M., Chaing, S., Rabolt, J. F., Swalen, J. D., and Wilson, R. J., *Phys. Rev. Lett.*, **59** 2879–82 (1987).
3. Hoffmann-Millack, B., Roberts, C. J., and Steer, W. S., *J. Appl. Phys.*, **67**, 1749–52 (1990).
4. Wintterlin, J., Wiechers, J., Burne, H., Gritsch, T., Höfer, H., and Behm, R. J., *Phys. Rev. Lett.*, **62**, 59–62 (1989).
5. Lippel, P. H., Wilson, R. J., Miller, M. D., Wöll, Ch., and Chaing, S., *Phys. Rev. Lett.*, **62**, 171–4 (1989).
6. Tromp, R. M., Hammers, R. J., and Demuth, J. E., *Phys. Rev. B*, **34** 1388–91 (1988).
7. Eigler, D. M., and Schweizer, E. K., *Nature*, **344**, 524–6 (1990).
8. Baró, A. M., Miranda, R., and Carrascosa, J. L., *IBM J. Res. Develop.*, **30**(4), 380–6 (1986).
9. Sonnenfeld, P., and Hansma, P. K., *Science*, **232**, 211–13 (1986).
10. Hansma, P. K., Elings, V. B., Marti, O., and Bracker, C. E., *Science*, **242**, 209–16 (1988).
11. Salmeron, M., Beebe, T., Odriozola, J., Wilson, T., Ogletree, D. F., and Siekhaus, W., *J. Vac. Sci. Technol.*, **A8**(1), 635–41 (1990).
12. Simmons, J. G., *J. Appl. Phys.*, **34** 1793–803 (1963).
13. Tersoff, J., *Phys. Rev. B*, **40**, 11900–3 (1989); **41**, 1235–8 (1990).
14. Lang, N. D., *Phys. Rev. Lett.*, **55**, 230–3 (1985); **56**, 1164–7 (1986).
15. Sacks, W., and Noguera, C., *J. Vac. Sci. Technol.*, **B9**(2), 488–91 (1991).
16. Hansma, P. K., and Tersoff, J., *J. Appl. Phys.*, **61**(2), R1–R23 (1987).
17. Rabe, J. P., and Bucholz, S., *Science*, **253** 424–6 (1991).
18. Laegsgard, E., Besenbacher, F., Mortensen, K., and Stensgard, I., *Rev. Sci. Instrum.*, **59**, 1035–8 (1986).
19. Demuth, J. E., Hammers, R. J., Tromp, R. M., and Welland, M. E., *IBM J. Res. Develop.*, **30**(4), 396–402 (1986).
20. Albrecht, O., Madsen, L. L., Mygind, J., and Mørch, K. A., *J. Phys. E: Sci. Instrum.*, **22**, 39–42 (1989).
21. V. G. Microtech Ltd, Bellbrock Business Park, Bell Lane, Uckfield, East Sussex, TN22 1QZ.
22. Digital Instruments Inc., 6780 Cortona Drive, Santa Barbara, California 93177, USA.
23. Park Scientific Instruments, 476 Ellis Street, Mountain View, California 04043, USA.
24. Binnig, G., and Rohrer, H., in J. Janka and J. Pantoflicek (eds.), *Trends in Physics*, European Physics Society, The Hague (1984), pp. 38–46.
25. Baró, A. M., Miranda, R., Alaman, J., García, N., Binnig, G., Rohrer, H., Gerber, Ch., and Carracosa, J. L., *Nature*, **315** 253–4 (1985).
26. Snellman, H., Pelliniemi, L. J., Penttinen, R., and Laiho, R., *J. Vac. Sci. Technol.*, **A8**(1), 692–4 (1990).
27. McMaster, T. J., Carr, H., Miles, M. J., Cairns, P., and Morris, V. J., *J. Vac. Sci. Technol.*, **A8**(1), 648–50 (1990).
28. Elings, V. B., Edtstrom, R. D., Meinke, M. H., Yang, X., Yang, R., and Evans, D. F., *J. Vac. Sci. Technol.*, **A8**(1), 652–4 (1990).
29. Shigekewa, H., Morozumi, T., Komiyama, M., Yoshimura, M., Kawazu, A., and Saito, Y., *J. Vac. Sci. Technol.*, **B9**(2), 1189–92 (1991).
30. Miles, M. J., Lee, I., and Atkins, E. D. T., *J. Vac. Sci. Technol.*, **B9**(2), 1206–9 (1991).
31. Arscott, P. G., Lee, G., Bloomfield, V. A., and Evans, D. F., *Nature*, **339** 484–6 (1989).

32. De Stasio, G., Rioux, D., Margaritondo, G., Mercanti, D., Trasatti, L., and Moore, C., *J. Vac. Sci. Technol.*, **A9**(4), 2319–21 (1991).
33. Allison, D. P., Thompson, J. R., Jacobson, K. B., Warmack, R. J., and Ferrell, T. L., *Scan. Microsc.*, **4**(3), 517–22 (1990).
34. Amrein, M., Stasiak, A., Gross, H., Stoll, E., and Travaglini, G., *Science*, **240**, 514–16 (1988).
35. Amrein, M., Dürr, R., Stasiak, A., Gross, H., and Travaglini, G., *Science*, **243**, 1708–11 (1989).
36. Lindsay, S. M., and Barris, B., *J. Vac. Sci. Technol.*, **A6**(2) 544–7 (1988).
37. DeRose, J. A., Lindsay, S. M., Nagahara, L. A., Odin, P. I., Thundat, T., and Rill, R. L., *J. Vac. Sci. Technol.*, **B9**(2), 1166–70 (1991).
38. Rabe, J. P., Sano, M., Batchelder, D., and Kalatchev, A. A., *J. Microsc.*, **152**(2), 573–83 (1988).
39. Albrecht, T. R., Dovek, M. M., Lang, C. A., Grütter, P., Quate, C. F., Kuan, S. W. J., Frank, C. W., and Pease, R. F. W., *J. Appl. Phys.*, **64**(3), 1178–85 (1988).
40. Reneker, D. H., Schneir, J., Howell, B., and Harry, H., *Polym. Commun.*, **31** 167–9 (1990).
41. Dovek, M. M., Albrecht, T. R., Kuas, S. W. J., Lang, C. A., Emch, R., Grütter, P., Frank, C. W., Pease, R. F. W., and Quate, C. F., *J. Microsc.*, **152**(1), 229–36 (1988).
42. Roberts, C. J., Sekowski, M., Davies, M. C., Jackson, D. E., Price, M., and Tendler, S. J. B., *Biochem. J.*, **283**, 181 (1992).
43. Driscoll, R. J., Youngquist, M. G., and Baldeschieler, J. D., *Nature*, **346** 294–6 (1990).
44. Keller, R. W., Dunlap, D. D., Keller, D. J., Garcia, R. G., Gray, C., and Maestre, M. F., *J. Vac. Sci. Technol.*, **A8**(1), 706–12 (1990).
45. Smith, D. P. E., Hörber, J. K. H., Binnig, G., and Nejoh, H., *Nature*, **344**, 641–4 (1990).
46. Hara, M., Iwakabe, Y., Tochig, K., Sasake, H., Garito, A. F., and Yamada, Y., *Nature*, **344**, 228–30 (1990).
47. Heckl, W. M., Kallury, K. M. R., Thompson, M., Gerber, Ch., Hörber, H. J. K., and Binnig, G., *Langmuir*, **5**, 1433–5 (1989).
48. Stemmer, A., Reichelt, R., Engel, A., Rosenbusch, J. P., Ringger, M., Hidber, H. R., and Güntherodt, H.-J., *Surf. Sci.*, **181**, 394–402 (1987).
49. Salmeron, M., Beebe, T., Ordiozola, J., Wilson, T., Ogletree, D. F., and Siekhaus, W., *J. Vac. Sci. Technol.*, **A8**(1), 635–41 (1990).
50. Lindsay, S. M., Sankey, O. F., Li, Y., Herbst, C., and Rupperecht, A., *J. Phys. Chem.*, **94**, 4655–60 (1990).
51. García, R., and García, N., *Chem. Phys. Lett.*, **173**(1), 44–50 (1990).
52. Stemmer, A., and Engel, A., *Ultramicrosc.*, **34**, 129–40 (1990).
53. Andrade, J. D. (ed.), *Surface and Interfacial Aspects of Biomedical Polymers*, Plenum Press, New York (1985).
54. Roberts, C. J., Wilkins, M. J., Davies, M. C., Jackson, D. E., and Tendler, S. J. B., *Surf. Sci. Lett.*, **261**(1–3), L29 (1992).
55. Wilson, T. E., Murray, M. N., Ogletree, D. F., Bednarski, M. D., Cantor, C. R., and Salmeron, M. B., *J. Vac. Sci. Technol.*, **B9**(2), 1171–6 (1991).
56. Wilkins, M. J., Roberts, C. J., Davies, M. C., Jackson, D. E., and Tendler, S. J. B., in *Scanning tunnelling microscopy: a proposed way Forward*, Batts, G. and West, R. eds, Butterworth, London, in press.
57. Olk, C. H., Heremann, J., Lee, P. S., Dziedzic, D., and Sargent, M., *J. Vac. Sci. Technol.*, **B9**(2), 1268–71 (1991).
58. Cricenti, A., Selci, S., Chiarotti, G., and Amaldi, F., *J. Vac. Sci. Technol.*, **B9**(2), 1285–7 (1991).
59. Cricenti, A., Selci, S., Felici, A. C., Gemerosi, R., Gori, E., Djaczenko, W., and Chiarotti, G., *Science*, **245**, 1226–7 (1989).

60. Foster, J. S., Frommer, J. E., and Arnett, P. C., *Nature*, **331**, 324–6 (1988).
61. Eigler, D. M., and Sweizer, E. K., *Nature*, **344**, 524–6 (1990).
62. Mamin, H. J., Guenthner, P. H., and Rugar, D., *Phys. Rev. Lett.*, **65**, 2418–21 (1990).
63. Allen, M. J., Tench, R. J., Mazrimas, J. A., Balooch, M., Sickhaus, W. J., and Balhom, R., *J. Vac. Sci. Technol.*, **B9**(2), 1272–5 (1991).
64. Wilkins, M. J., Roberts, C. J., Davies, M. C., Jackson, D. E., and Tendler, S. J. B., *Ultramicroscopy*, in press.
65. Clemmer, C. R., and Beebe Jnr, T. P., *Science*, **251**, 640–2 (1991).
66. Roberts, C. J., Wilkins, M. J., Beamson, G., Davies, M. C., Jackson, D. E., Tendler, S. J. B., and Williams, P. M., *Nanotechnology*, **3**(2), 98 (1992).
67. Nawaz, Z., Cataldi, T. R. I., Knall, J., Somekh, R., and Pethica, J. B., *Surf. Sci.*, 139 (1992).
68. Heck, W. M., and Binnig, G., *Ultramicroscopy*, **42–44**, 1073 (1992).
69. Edelman, V. S., Troyanovskii, A. M., Khaikin, M. S., Stepanyan, G. A., and Volodin, A. P., *J. Vac. Sci. Technol.*, **B9**(2), 618–22 (1991).
70. Roberts, C. J., Davies, M. C., Jackson, D. E., Tendler, S. J. B., and Williams, P. M., *J. Phys.: Condens. Matter*, **3**, 7213–16 (1991).
71. Mizutani, W., Inukai, J., and Ono, M., *Japan J. Appl. Phys.*, **29**, L815–L817 (1990).
72. Rabe, J. P., Bucholz, S., and Ritchey, A. M., *J. Vac. Sci. Technol.*, **A8**(1), 679–83 (1990).
73. Binnig, G., Quate, C. F., and Gerber, Ch., *Phys. Rev. Lett.*, **56**, 930–3 (1986).
74. Drake, B., Prater, C. B., Weisenhorn, A. L., Gould, S. A. C., and Hansma, P. K., *Science*, **244**, 1586–9 (1989).
75. Häberle, W., Hörber, J. K. H., and Binnig, G., *J. Vac. Sci. Technol.*, **B9**(2), 1210–13 (1991).
76. Sarid, D., and Elings, V., *J. Vac. Sci Technol.*, **B9**(2), 431–7 (1991).
77. Binnig, G., Quate, C. F., and Gerber, Ch., *Phys. Rev. Lett.*, **56**, 930–3 (1986).
78. Meyer, E., Heinzelmann, H., Rudin, H., and Güntherodt, H.-J., *Z. Phys. B.—Condens. Matter*, **79**, 3–4 (1990).
79. Ducker, W. A., Senden, T. J., and Pashley, R. M., *Nature*, **353**, 239–41 (1991), and Gould, S. A. C., Drake, B., Prater, C. B., Weisenhorn, A. L., Manne, S., Hansma, H. G., Massie, J., Longmire, M., Elings, V. E., Northern, B., Mukergee, B., Peteson, C. M., Stoeckesieus, W., Albrecht, T. R., and Quate, C. F., *J. Vac. Sci. Technol.*, **A8**(1), 369–73 (1990).
80. Gould, S., Marti, O., Drake, B., Hellemans, L., Bracker, C. E., Hansma, P. K., Keder, N. L., Eddy, M. M., and Stucky, G. D., *Nature*, **332**, 332–4 (1988).
81. Butt, H.-J., Prater, C. B., and Hansma, P. K., *J. Vac. Sci. Technol.*, **B9**(2), 1193–6 (1991).
82. Drake, B., Prater, C. B., Weisenhorn, A. L., Gould, S. A. C., Albrecht, T. R., Quate, C. F., Cannell, D. S., Hansma, H. G., and Hansma, P. K., *Science*, **243**, 1586–9 (1989).
83. Hoh, J. H., Lal, R., John, S. A., Revel, J.-P., and Arnsdorf, M. F., *Science*, **253**, 1405–7 (1991).
84. Dürig, U., Züger, O., and Pohl, D. W., *J. Microsc.*, **152**(1), 259–63 (1988).
85. Paesler, M. A., Moyer, P. J., Jahncke, C. J., Johnson, C. E., Reddick, R. C., Warmack, R. J., and Ferrel, T. L., *Phys. Rev. B*, **42**, 6750–3 (1990).
86. Williams, C. C., and Wickramasinghe, H. K., *Appl. Phys. Lett.*, **49**, 1587–89; and *Scan. Microsc.*, **2**, 3–8 (1988).
87. Williams, C. C., and Wickramasinghe, H. K., *Nature*, **344**, 317–19 (1990).

10

Surface Grafting of a Thrombin Substrate on a Polymer Membrane and the Inhibition of Thrombin Activity Leading to Non-thrombogenicity

Yoshihiro Ito, Lin-Shu Liu and Yukio Imanishi

Department of Polymer Chemistry,
Kyoto University

1 INTRODUCTION

A number of investigations have been made to develop non-thrombogenic materials by suppressing platelet activation,[1] activating the fibrinolytic system[2] and heparinization.[3] The main role of heparin is to inhibit the catalytic action of thrombin and thus to suppress activation of the blood coagulation system. Thrombin accelerates the conversion of fibrinogen to fibrin, and this process is inhibited by heparin through activation of antithrombin III.

In the present investigation, a series of new derivatives of thrombin substrate were synthesized and immobilized on polymer membranes, in order to investigate the relation between the inhibitory effect on thrombin activity and non-thrombogenicity and to develop new non-thrombogenic materials. It has been shown that thrombin catalyses the hydrolytic scission of coagulation factor I (fibrinogen),[4] VII,[5] VIII[6] or XIII[7] at the carboxyl end of an arginine (Arg) residue adjacent to hydrophobic α-amino acid residues. It occurred to us that hydrophobic short-chain peptides containing an Arg residue might compete with the coagulation factors I (fibrinogen), VII, VIII and XIII for inhibition of the formation of activated complex in the hydrolytic reactions or the product might induce inhibition of the hydrolytic reactions. A previous investigation[8] of the reactivity of a number of peptide substrates in the thrombin-catalysed hydrolysis has shown that peptides containing the valine (Val)–proline (Pro)–arginine (Arg) sequence were most reactive in this respect.

Several peptides having a Val–Pro–Arg sequence were synthesized as thrombin substrate and immobilized on polyurethane membranes to investigate their

Polymer Surfaces and Interfaces II
Edited by W. J. Feast, H. S. Munro and R. W. Richards
© 1993 John Wiley & Sons Ltd

inhibitory effect on thrombin activity. The interactions with platelets and the *in vitro* non-thrombogenicity of the thrombin substrate immobilized materials were investigated. Furthermore, the thrombin catalysis at the solid–liquid interface was investigated by using synthetic fluorescent substrates immobilized on a polymer membrane. The immobilization of a synthetic fluorescent substrate on the surface of a polymer membrane should be useful for the quantitative investigation of thrombin catalysis on the insoluble matrix with respect to development of new blood-compatible materials.

2 EXPERIMENTAL

2.1 Thrombin Substrate Immobilized Polymer Membrane— Non-thrombogenic Membrane

The synthetic thrombin substrates, valyl-prolyl-arginyl-glycine (VPRG), valyl-prolyl-arginine methyl ester (VPR–OMe) and valyl-prolyl-arginine (VPR), were synthesized by the liquid-phase method as illustrated in Figure 1. The *t*-butoxycarbonyl (Boc) protecting groups of these synthetic peptides were removed with 4N HCl/dioxane. The purity of the product peptide was checked by elemental analysis.

To synthesize non-thrombogenic polyurethane membrane, the peptides were dissolved in water to various concentrations and immobilized on a poly(acrylic acid)-grafted polyetherurethane (PEU–PAA) membrane with a water-soluble carbodiimide (WSC). Polyetherurethane (PEU) was synthesized by chain extension with 1,4-dihydroxybutane of a urethane prepolymer, which was synthesized by the reaction of polyoxytetramethylene (POTM, molecular weight 870) and 4,4′-diisocyanate diphenylmethane (MDI) (1 : 2 mol/mol) in dimethylformamide

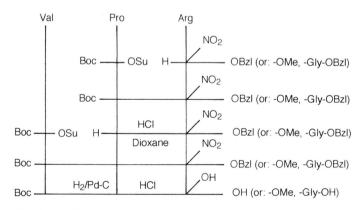

Figure 1. Synthesis of thrombin substrates

(DMF) solution. The PEU was dissolved in DMF (10 wt %) and cast as a film on a glass plate under irradiation with an infrared lamp.

The surface of the polymer membrane was treated by glow discharge (5 A, 10 min) with Eiko ion coater IB3 (Eiko Engineering Co. Ltd). The glow-discharged PEU membrane was immersed in an aqueous acrylic acid solution (10 wt %) under ultraviolet irradiation for 24 h to obtain the PEU–PAA membrane. The carboxyl groups introduced were determined by the method reported previously.[9] The amount of grafted poly(acrylic acid) (PAA) was found to be 1.95 μmol/cm^2.

The immobilization of the synthetic peptides on the PEU–PAA membrane was carried out under varying pH conditions, peptide concentrations and reaction times to find the optimum conditions for the coupling reaction between the terminal amino groups of the synthetic thrombin substrates and the carboxyl groups of the PEU–PAA membrane. The immobilized products are represented by PEU–PAA–P, in which P represents the synthetic peptides VPR, VPR–OMe, VPRG and so on.

The peptide-immobilized PEU–PAA membrane was heated at 110°C for 12 h in 6 N hydrochloric acid, and the solution was neutralized with 6 N aqueous NaOH solution. 4-Phenylspiro[furan-2(3H),1'-phthalan]-3,3'-dione (Fluram) was reacted with α-amino acids produced in the hydrolysis and the fluorescence intensity of the reaction product was measured with an excitation wavelength at 390 nm and an emission wavelength at 480 nm.

2.2 Fluorescent Thrombin Substrate Immobilized Polymer Membrane— Kinetic Investigations

The synthetic fluorescent substrates for thrombin catalysis, *t*-butoxycar-bonyl–valyl-prolyl-arginyl 4-methylcoumaryl-7-amide(Boc–VPR–MCA), *t*-butoxycarbonyl–isoleucyl-glutamyl-glycyl-arginyl 4-methylcoumaryl-7-amide (Boc–IEGR–MCA), prolyl-phenylalanyl-arginyl 4-methylcoumaryl-7-amide (PFR–MCA) and isoleucyl-prolyl-arginyl *p*-nitroanilide (IPR–PNA) were purchased from Nacalai Tesque Inc. and used for kinetic investigations of thrombin catalysis. Fibrinogen (from bovine plasma, F-8630) and thrombin (from bovine plasma, T-4265) were purchased from Sigma Chemical Co.

The polyetherurethaneurea (PEUU) membrane with carboxylic groups on the surface was synthesized as previously reported.[10] The polymer was cast into a membrane (1.77 cm^2) on a glass plate from a DMF solution (10 wt %). The film was immersed in a mixture of 4N aqueous citric acid solution (10 wt %) and methanol (1 : 3 v/v) for 1 h at room temperature. Next, 6-aminocaproic acid as a spacer chain was connected to the carbonyl groups on the membrane surface. The amount of the spacer chain introduced was determined using Rhodamine 6G as previously reported.[10] It was found that the amount of carboxyl groups used in the reaction on the surface was 32%.

Boc–VPR–MCA and Boc–IEGR–MCA were deblocked by treating with 4N HCl/dioxane. Immobilization of the synthetic thrombin substrate was performed in 2-(N-morpholino)ethanesulphonic acid (MES) buffered solution (pH 4.7) containing WSC (10 wt %). The membranes, on which VPR–MCA, IEGR–MCA and PFR–MCA were immobilized, will be referred to as PEUU–VPR–MCA, PEUU–IEGR–MCA and PEUU–PFR–MCA, respectively.

The amount of immobilized substrates was determined by the hydrolysis of the membrane in 6 N HCl (5 ml) for 6 h at 110°C. After the reaction, the solution was neutralized with 6 N NaOH and diluted with tris-HCl buffered solution (pH 8) to a total volume of 150 ml. Fluorescent intensity at 460 nm was measured with the excitation at 380 nm. A calibration curve was obtained by using 7-amino-4-methylcoumarin (AMC).

3 RESULTS AND DISCUSSION

3.1 Non-thrombogenic Polyurethane Membrane

3.1.1 Inhibition of Thrombin Activity by the Synthetic Substrates

The inhibition of thrombin activity by the synthetic substrates was assessed by determining the amount of fibrinopeptide formed according to the method reported by Bando et al.[11] The rate of fibrinopeptide formation was slowed down by the addition of the synthetic substrates as shown in Figure 2. The suppressive effect of the synthetic substrates increased in the order of VPR < VPRG < VPR–OMe. The lower inhibition by VPR could be explained in terms of its lower competitivity with fibrinogen towards the binding site of the enzyme. The inhibitory effects on thrombin activity of the synthetic substrates were compared under similar conditions and are shown in Table 1. It was found that the inhibitory effects of the synthetic substrates increased with increasing chain length in the three series of peptide analogues, i.e. VPR > PR > Boc–R; VPR–OMe > PR–OMe; VPRG > PRG > RG. This result is probably a reflection of specific interaction in the reaction of longer chain substrates.

3.1.2 Immobilization of the Synthetic Substrates

The dependence on reaction conditions of the amount of the synthetic substrates immobilized on the PEU–PAA membrane surface was investigated under various conditions. The maximum immobilization was attained in a 16 h reaction at room temperature and pH 4.7 and at room temperature with a peptide concentration above 8 wt %.

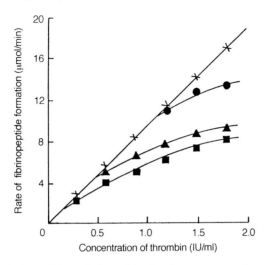

Figure 2. Proteolytic activity of thrombin in the presence of synthetic thrombin substrates or fibrinogen (8 μM) at pH 7.0, 37°C: X fibrinogen, ● VPR, ▲ VPRG, ■ VPR–OMe

The amounts of the synthetic substrates immobilized on to the PEU–PAA membrane in the reaction at pH 4.7 at room temperature for 24 h with a concentration of 10 wt % are shown in Table 2. Under these conditions, 1.5–2.4 mol % of the synthetic peptides added to the reaction solution were immobilized on the PEU–PAA film. This amount of immobilization corresponds to the usage of 14–35 mol % of carboxyl groups of the PAA grafts for coupling with the terminal amino groups of the synthetic substrates.

Table 1. Effect of synthetic substrate on the rate of fibrinopeptide formation[a]

Synthetic substrate	Rate (μmol/min)
VPR	15.1
PR	16.0
Boc–R	18.5
VPR–OMe	8.5
PR–OMe	9.6
VPRG	9.0
PRG	10.0
RG	15.7

[a]Fibrinogen 8.0 μM, thrombin 2.0 IU/ml, synthetic substrate 8.0 μM at 37°C in citrate buffer (pH 7.0).

Table 2. Characterization of PEU–PAA films immobilized with various synthetic substrates

Film	Amount of immobilization ($\times 10^7$ mol/cm^2)	Contact angle (deg)	Relative amount of thrombus formed (%)
Glass	—	32.5 ± 1.5	100
PEU	—	55.0 ± 1.0	52 ± 2.0
PEU–PAA	—	39.0 ± 1.0	49 ± 1.5
PEU–PAA–VPR	5.36 ± 0.21	35.5 ± 1.5	29 ± 1.5
PEU–PAA–PR	5.42 ± 0.15	35.5 ± 1.5	28 ± 1.0
PEU–PAA–R	7.21 ± 0.13	35.5 ± 2.0	28 ± 5.5
PEU–PAA–VPR–OMe	4.89 ± 0.17	36.5 ± 1.0	8 ± 2.5
PEU–PAA–PR–OMe	4.71 ± 0.16	37.0 ± 1.5	10 ± 2.5
PEU–PAA–VPRG	5.01 ± 0.14	37.0 ± 1.0	10 ± 2.0
PEU–PAA–PRG	5.33 ± 0.20	36.5 ± 1.0	12 ± 1.5
PEU–PAA–RG	5.74 ± 0.19	36.5 ± 0.5	27 ± 4.5

3.1.3 Surface Properties of the Polymer Membranes

The contact angles of an air bubble in water on the PEU–PAA–P membranes are shown in Table 2. The grafting of PAA chains on to the PEU membrane strongly increased the wettability of the membrane. However, the immobilization of the synthetic peptides on to the PEU–PAA membrane increased the wettability of the membrane only slightly. The wettability of the PEU–PAA–P membranes was not clearly dependent on the amount and the nature of the synthetic substrates immobilized.

3.1.4 Interactions with Thrombin of the Thrombin Substrates Immobilized on to the Polymer Membranes

The effects of the polymer membranes immobilized with various synthetic substrates on the hydrolytic reaction of fibrinogen catalysed by thrombin are shown in Figures 3 and 4. The rate of fibrinopeptide formation increased with increasing concentration of fibrinogen, but it was slowed down by the addition of the polymer membranes immobilized with the synthetic substrates. As will be described in Sections 3.2.1 and 3.2.3, the PEU membrane interferes with thrombin activity. However, the order of the inhibitory effects of the synthetic substrates immobilized on the polymer membranes was the same as that of free synthetic substrates. This finding indicates that the inhibition of fibrinopeptide formation is mainly based on the specific interactions between the synthetic substrates and thrombin. However, the overall inhibiting effects of the immobilized substrates were lower than those of free substrates. This is evident by

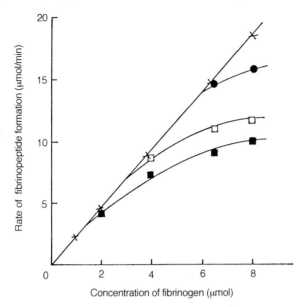

Figure 3. Proteolytic activity of thrombin (2.0 IU/ml) in the presence of immobilized thrombin substrates (8.0 μM) and different concentrations of fibrinogen at pH 7.0, 37°C: X fibrinogen (F) only, ● PEU–PAA–VPR and F, ■ PEU–PAA–VPR–OMe and F, □ PEU–PAA–PR–OMe and F

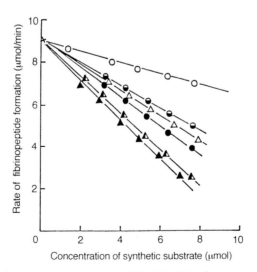

Figure 4. Proteolytic activity of thrombin (2.0 IU/ml) in the presence of immobilized thrombin substrates of different concentrations at pH 7.0, 37°C, the concentration of fibrinogen being 2.5 μM: ○ PEU–PAA–R, ◍ PEU–PAA–PR, ● PEU–PAA–VPR, △ PEU–PAA–RG, △ PEU–PAA–PRG, ▲ PEU–PAA–VPRG

comparing the amounts of fibrinopeptides formed in Figures 2 and 3. Under conditions where the concentration of thrombin is 2.0 IU/ml and that of synthetic substrate 8 μM, the amounts of fibrinopeptide formed in the presence of VPR, VPRG and VPR–OMe were 13.6, 9.4 and 8.3 μmol, respectively (Figure 2). When the immobilized substrates were used, they were 15.8, 11.4 and 9.67 μmol, respectively (Figure 3). The decreased reactivity of the immobilized substrates may be explained by confinement of the thrombin–synthetic peptide interaction into a two-dimensional plane. A deterioration of thrombin activity due to non-specific interaction with the surface of the immobilization matrix should also be taken into account as a reason for the apparently low activity of immobilized substrates, as will be described in Sections 3.2.1 and 3.2.3.

3.1.5 Determination of the Amount and the Activity of Thrombin Adsorbed on the Membrane Surface

The PEU–PAA–P membrane was immersed in a phosphate buffered saline (PBS, pH 7.0, 1 ml) containing thrombin (1.0 mg/ml) at 37°C under stirring for a requisite time. The membrane was washed three times in PBS for a minute at 37°C under stirring until no thrombin was released from the membrane to PBS, as detected by a ninhydrin test.

The thrombin-adsorbed membrane was put in distilled water containing sodium dodecylsulphate (2 wt %) and subjected to ultrasonic treatment twice at room temperature for 30 min each. The washing liquid was collected, condensed to about 1 ml and added to concentrated hydrochloric acid (1 ml). The acidic solution was refluxed at 110°C for 12 h and neutralized with 6 N aqueous NaOH solution. The amount of thrombin adsorbed was determined by the method of α-amino acid determination, as described above.

The thrombin-adsorbed membrane, which was washed as described above, was equilibrated at 37°C for 3 min and transferred to PBS (1.0 ml) containing Boc–Val–Pro–Arg–MCA (0.1 mM). The reaction was continued for 10 min at 37°C under stirring, and the thrombin activity was assessed on the basis of the amount of AMC liberated by the thrombin catalysis.

The time-dependent thrombin adsorption on to the PEU–PAA–P membranes is shown in Figure 5. The same amount of thrombin adsorbed was attained on any PEU–PAA–P membrane by incubation of the polymer membrane in an aqueous thrombin solution for a longer time than 100 min. However, the rate of thrombin adsorption was strongly dependent on the nature of the synthetic substrate immobilized, and the time required for reaching the saturation amount of adsorption was shortened by the immobilization of peptides with increasing inhibitory effects on the thrombin activity. The activity of adsorbed thrombin was lower when the PEU–PAA–P membranes having stronger inhibitory effects were used (see Table 3).

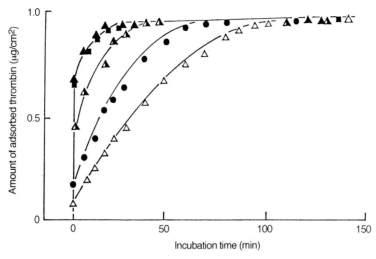

Figure 5. Adsorption of thrombin (1 mg/ml) to PEU–PAA membrane immobilized with synthetic substrates in PBS at 4°C: ▲ PEU–PAA–VPRG, △ PEU–PAA–PRG, △ PEU–PAA–RG, ■ PEU–PAA–VPR–OMe, ● PEU–PAA–VPR

3.1.6   Non-thrombogenicity and Interactions with Platelets of the Polymer Membranes with Immobilized Synthetic Substrates

The adhesion and activation of platelets on the polymer membranes synthesized in the present investigation were determined by the method reported previously.[12] The number of platelets adhering to the PEU film was decreased by

Table 3. Amount and activity of thrombin adsorbed on the film surface immobilized with various synthetic substrates[a]

Film	Amount of adsorbed thrombin (μg/cm^2)	Activity of adsorbed thrombin (%)
PEU	0.95 ± 0.03	94 ± 2
PEU–PAA	0.93 ± 0.04	95 ± 2
PEU–PAA–VPR	0.92 ± 0.05	64 ± 3
PEU–PAA–PR	0.96 ± 0.02	68 ± 4
PEU–PAA–R	0.94 ± 0.04	84 ± 4
PEU–PAA–VPR–OMe	0.94 ± 0.03	5 ± 1
PEU–PAA–PR–OMe	0.92 ± 0.04	8 ± 2
PEU–PAA–VPRG	0.92 ± 0.03	5 ± 2
PEU–PAA–PRG	0.96 ± 0.05	10 ± 2
PEU–PAA–RG	0.93 ± 0.04	66 ± 4

[a] Incubation at 4°C in PBS for 90 min with thrombin concentration 1.0 mg/ml.

Table 4. Number of platelets attached to the film surface immobilized with various thrombin substrates

Film	Radioactivity of adhered platelets (cpm)
PEU	693 ± 30
PEU–PAA	565 ± 13
PEU–PAA–VPR	439 ± 18
PEU–PAA–PR	448 ± 14
PEU–PAA–R	517 ± 19
PEU–PAA–VPR–OMe	217 ± 17
PEU–PAA–VPRG	248 ± 14
PEU–PAA–RG	403 ± 20

grafting PAA chains (Table 4) and further decreased by immobilization of the synthetic substrates. The number of platelets adhering to the PEU–PAA–P membranes decreased when the peptides having stronger inhibition effects were immobilized (Table 4).

Figure 6 shows the SEM photographs of the platelets adhering to different polymer membranes. When adhered to polymer membranes, the platelets are deformed and the formation of pseudo pods is observed. The suppression of platelet adhesion on the surface by immobilized thrombin substrate peptides is due either to prevention of fibrin network formation or to inhibition of thrombin activity towards platelet activation.[13,14] The amount of thrombus formed on the polymer membranes was determined by the method reported previously,[9] using the blood collected from an adult dog to estimate the *in vitro* non-thrombogenicity of the polymer membranes prepared. The amount of thrombus formed in the *in vitro* clotting test is shown in Table 2. The thrombus formation was suppressed slightly by grafting PAA chains on to the PEU membrane. It could be ascribed to the increasing wettability of the polymer membrane. The thrombus formation was strongly suppressed by immobilization of the synthetic substrates. The decreasing thrombus formation is probably due to suppression of the fibrinogen/fibrin conversion, the polymerization of monomeric fibrin and the activation platelet. The thrombus formation was suppressed more effectively by higher amounts of immobilization the synthetic substrates or by immobilization of the synthetic substrates having higher inhibitory effects.

By comparison with literature values it was found that the thrombin substrate immobilized PEU membrane is similarly non-thrombogenic to poly(sodium vinyl sulphonate) immobilized[15] and urokinase immobilized[2] polyurethane membranes, and is definitely more non-thrombogenic than polyurethane or PEU–PAA membrane.

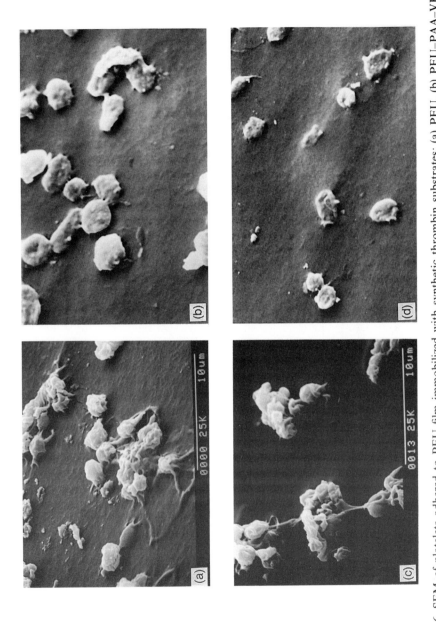

Figure 6. SEM of platelets adhered to PEU film immobilized with synthetic thrombin substrates: (a) PEU, (b) PEU–PAA–VPR, (c) PEU–PAA–R, (d) PEU–PAA–VPRG

3.2 Kinetic Investigation of Thrombin Catalysis

3.2.1 Hydrolysis of Fluorescent Thrombin Substrate by Thrombin

Thrombin in PBS (4.0 μg/ml, 50 μl) was added to tris-HCl solution (2.5 ml) containing free or immobilized synthetic substrate at 37°C under stirring. The amount of released AMC was determined by fluorescence intensity. When IPR–PNA was used as a substrate, the reaction rate was determined by measuring the ultraviolet absorption of *p*-nitroaniline liberated.

The Lineweaver–Burk plot of the rate of hydrolysis of synthetic substrates with thrombin gave the dissociation constant (K) and the catalytic constant (k)

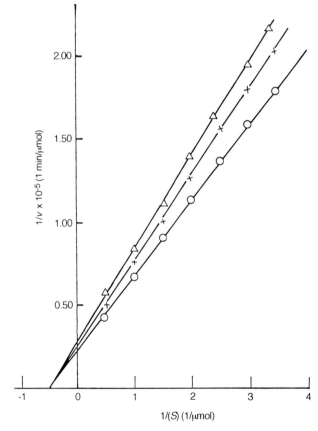

Figure 7. Lineweaver–Burk plot of the hydrolysis of Boc–VPR–MCA catalysed by thrombin in the absence or presence of PEUU membranes at 37°C in PBS (1 ml). The concentration of thrombin was 2.2 μg/ml: $\bigcirc$ in the absence of PEUU membrane, X in the presence of PEUU membrane (1 cm^2), $\triangle$ in the presence of PEUU membrane (2 cm^2)

Table 5. Kinetic parameters for the hydrolysis of free and immobilized thrombin substrates

Boc–VPR–MCA	2	34	1.7×10^{-1}
Boc–IEGR–MCA	33	10	3.0×10^{-3}
PFR–MCA	63	6.9	1.1×10^{-3}
IPR–PNA	3	29	9.7×10^{-2}
PEUU–VPR–MCA	27	24	8.9×10^{-3}
PEUU–IGER–MCA	270	4.7	1.7×10^{-4}
PEUU–PFR–MCA	526	3.8	7.2×10^{-5}
Fibrinogen	12	—	—

that are shown in Table 5. The K value increased in the order of Boc–VPR–MCA < Boc–IEGR–MCA < PFR–MCA, which agrees with the order reported by Morita *et al.*[8]

In Figure 7, the effect of PEUU membrane on the thrombin reaction with Boc–VPR–MCA is shown. It is found that the thrombin reaction is inhibited non-specifically by the membrane.

The immobilized substrates were hydrolysed by thrombin to liberate AMC molecules, as shown in Figure 8. It seems that the inhibition is a mixture of specific and non-specific activity and that the specific inhibition is more important than the non-specific inhibition. Assuming that non-specific interactions between thrombin and the sample membrane are negligible, K and k were

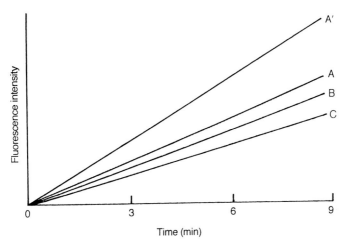

Figure 8. Time course of the release of AMC molecules from PEUU–VPR–MCA at 37°C in PBS (1 ml). The surface area of membrane was 1 cm² in A, B and C and 2 cm² in A'. The concentration of thrombin was 4.4 μg/ml in A and A', 2.2 μg/ml in B and 1.1 μg/ml in C

determined from the Lineweaver–Burk plot of the reaction and are shown in Table 5. The reactivities of the immobilized substrates were lower than those of free substrates, i.e. K increased and k decreased by immobilization. However, the order of the hydrolysis rate of the immobilized substrates was the same as that of the free substrates.

3.2.2 Catalytic Activity of Adsorbed Thrombin

Figure 9 shows the dependence of the catalytic activity in the hydrolysis of Boc–VPR–MCA of thrombin adsorbed on various substrate immobilized PEUU membranes on the amount of immobilized synthetic substrate. The activity of adsorbed thrombin was lowered with increasing amounts of immobilized substrate. Inactivation of thrombin by adsorption was more serious on a membrane surface on which a more reactive substrate was immobilized.

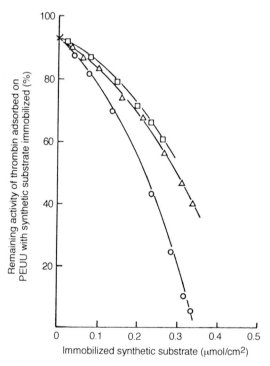

Figure 9. The catalytic activity of thrombin adsorbed on various PEUU membranes in the hydrolysis of Boc–VPR–MCA at 37°C in tris-HCl buffered solution in the presence of PEUU–VPR–MCA (○), PEUU–IEGR–MCA (△), PEUU–PFR–MCA (□) or PEUU (X)

3.2.3 Competitive Inhibition of Thrombin Catalysis by Synthetic Substrate in Fibrinogen Cleavage

The competitive inhibition of thrombin catalysis by synthetic substrate was assessed by determining the amount of fibrinopeptide formed according to the method described in Section 3.1.4. The experimental results on the hydrolytic reactions of IPR–PNA by thrombin in the presence of Boc–VPR–MCA or immobilized VPR–MCA are shown in Figure 10, along with the calculated values. The rate of the hydrolysis of IPR–PNA decreased with increasing concentration of Boc–VPR–MCA added. If the mechanism of the competing reactions is represented by Figure 11, the following equations are obtained:

$$v_1 = \frac{V^1_{max}[S_1]}{K_1 + [S_1] + K_1[S_2]/K_2} \tag{1}$$

$$v_2 = \frac{V^2_{max}[S_2]}{K_1 + [S_2] + K_2[S_1]/K_1} \tag{2}$$

where v_1 and v_2 represent the hydrolysis rate of the substrates S_1 and S_2, V^1_{max} and $^2_{max}$ their maximum rate of hydrolysis, $[S_1]$ and $[S_2]$ their concentration, and K_1 and K_2 the dissociation constant of the enzyme–substrate complexes, respectively.

The hydrolysis rate of IPR–PNA (S_1) in the presence of Boc–VPR–MCA (S_2) agreed roughly with the reaction rate v_1 calculated by using equation (1)

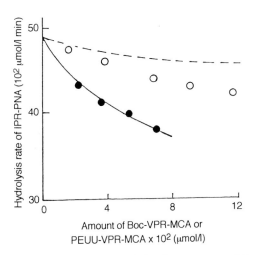

Figure 10. The hydrolysis rate of IPR–PNA in the presence of Boc–VPR–MCA (● observed value, —— calculated value by equation (1)) or PEUU–VPR–MCA (○ observed value, --- calculated value by equation (1))

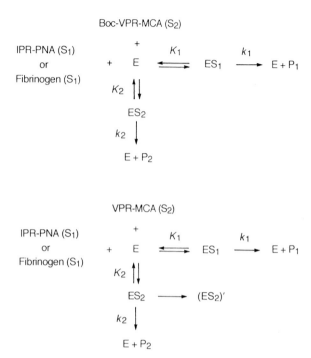

Figure 11. The mechanism of competitive inhibition of thrombin catalysis by synthetic substrate. E, thrombin; P_1, *p*-nitroaniline or fibrinopeptide; P_2, 7-amino-4-methyl-coumarin. (a) Thrombin catalysis in the presence of Boc–VPR–MCA; (b) thrombin catalysis in the presence of PEUU–VPR–MCA

and the kinetic parameters obtained in the reaction of IPR–PNA or Boc–VPR–MCA with thrombin.

The rate of the hydrolysis of IPR–PNA also decreased with increasing amounts of immobilized VPR–MCA added. However, the reaction profile was not explained by substituting the kinetic parameters for equation (1), which were obtained in the reaction of IPR–PNA or immobilized VPR–MCA with thrombin. The occurrence of non-specific inhibition by the PEUU membrane in the enzymatic reaction of Boc–VPR–MCA was suggested in Section 3.2.1. It may also occur in the enzymatic hydrolysis of IPR–PNA in the presence of PEUU membrane on which VPR–MCA is immobilized. Under these conditions, K_1 might be influenced by the presence of PEUU membrane, leading to a disagreement between the observed and calculated rates of IPR–PNA in the binary reaction system.

Similar observations were made in the inhibition of fibrinopeptide formation by free and immobilized synthetic thrombin substrate, as shown in Figures 12 and 13. More reactive synthetic substrate in the hydrolysis by thrombin

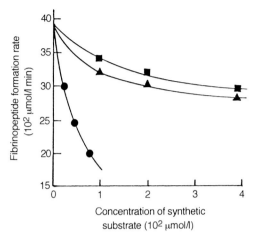

Figure 12. Fibrinopeptide formation by thrombin in the presence of Boc–VPR–MCA (●), Boc–IEGR–MCA (▲) or PFR–MCA (■). Real lines represent the reaction rate calculated by using equation (1)

retarded the cleavage of fibrinogen by thrombin more strongly. The hydrolysis rate of fibrinogen (S_1) in the presence of Boc–VPR–MCA (S_2) agreed roughly with the reaction rate v_1 calculated using equation (1) and the kinetic parameters obtained in the reaction of fibrinogen or Boc–VPR–MCA with thrombin. On the other hand, in the hydrolysis of fibrinogen by thrombin in the presence

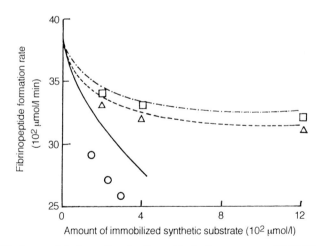

Figure 13. Fibrinopeptide formation by thrombin in the presence of PEUU–VPR–MCA (○ observed value, ——— calculated value by equation (1)), PEUU–IEGR–MCA (△ observed value, --- calculated value by equation (1)) or PEUU–PFR–MCA (□ observed value, -·-·- calculated value by equation (1))

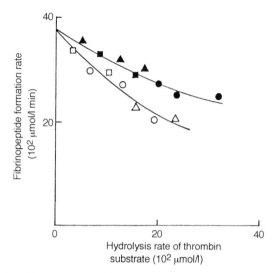

Figure 14. Relationship between inhibition of fibrinopeptide formation by synthetic substrates and the hydrolysis rate of the synthetic substrates: ● Boc–VPR–MCA, ▲ Boc–IEGR–MCA, ■ PFR–MCA, ○ PEUU–VPR–MCA, △ PEUU–IEGR–MCA, □ PEUU–PFR–MCA. The concentrations of synthetic substrates were varied. The inhibition and the hydrolysis are compared under the same concentrations of thrombin and the synthetic substrates

of immobilized substrates, the observed rate of hydrolysis did not agree with the calculated one. The non-specific inhibition by PEUU membrane in the thrombin catalysis on fibrinogen may explain the disagreement.

The rate of fibrinopeptide formation by thrombin in the presence of free or immobilized synthetic substrate is related to the rate of hydrolysis by thrombin of each synthetic substrate (free and immobilized) in Figure 14. More reactive synthetic substrate (with smaller K_2) in the hydrolysis by thrombin retarded the fibrinopeptide formation (v_1 decrease) by thrombin more strongly. For each synthetic substrate, a rise of concentration ($[S_2]$ rise) decreased the rate of fibrinopeptide formation. These inhibitory effects of synthetic substrate were stronger in the immobilized state than in solution, though the immobilization reduced the reactivity of synthetic substrate. This apparent discrepancy is explained in terms of the non-specific inhibition by PEUU membrane (increase of K_1 or decrease of k_1).

4 CONCLUSIONS

Synthetic thrombin substrates, which were immobilized on to PEUU or PEU membrane, were hydrolysed by thrombin. Although the reactivity in hydrolysis was reduced by immobilization, the rate of hydrolysis and the thrombin adsorption were still rapid on the membrane on which highly reactive substrate was immobilized. In addition, the catalytic activity of thrombin adsorbed on the membrane surface was strongly depressed. These results indicate the occurrence of specific interaction between thrombin and immobilized substrate. The increase of the Michaelis constant by immobilization should be due to the steric inhibition of substrate binding by thrombin on the surface of substrate immobilized membrane. The decreased catalytic constant should be a result of low reactivity, in other words high stability, of thrombin–substrate complex formed on the membrane surface. The stability of the thrombin–substrate complex on the membrane surface is reflected by the enhancement of thrombin adsorption (Table 3) and thrombin inactivation on adsorption (Figure 9). It was found that thrombin is bound non-specifically by the membrane surface as well as specifically by the immobilized substrate. The occurrence of specific and non-specific interactions was indicated by the competitive inhibition experiments.

The inhibition of thrombin catalysis accompanied inhibition of platelet adhesion. The present polyetherurethaneurea membrane on which thrombin substrate peptides are immobilized seems useful as a biodegradable blood-compatible biomaterial.

REFERENCES

1. Jacobs, H. A., Okano, T., and Kim, S. W., *J. Biomed. Mater. Res.*, **23**, 611 (1989).
2. Liu, L. S., Ito, Y., and Imanishi, Y., *Biomaterials*, **12**, 545 (1991).
3. Ito, Y., *J. Biomater. Appl.*, **2**, 235 (1987).
4. Henschen, A., Lottspeich, F., Kehl, M., and Southan, C., *Anal. N. Y. Acad. Sci.*, **408**, 28 (1983).
5. Radcliffe, R., and Nemerson, Y., *J. Biol. Chem.*, **251**, 4797 (1976).
6. Vehar, G. A., Keyt, B., Eaton, D., Rodriguez, H., O'Brien, D. P., Rotblat, F., Oppermann, H., Keck, R., Wood, W. I., Harkins, R. N., Tuddenham, E. G. D., Lawn, R. M., and Capon, D. J., *Nature*, **312**, 337 (1984).
7. Nakamura, S., Iwanaga, S., and Suzuki, T., *J. Biochem.*, **78**, 1247 (1975).
8. Morita, T., Kato, H., Iwanaga, S., Takada, K., Kimura, T., and Sakakibara, S., *J. Biochem.*, **82**, 1495 (1977).
9. Ito, Y., Sisido, M., and Imanishi, Y., *J. Biomed. Mater. Res.*, **20**, 1157 (1986).
10. Liu, L. S., Ito, Y., and Imanishi, Y., *Biomaterials*, **12**, 390 (1991).
11. Bando, M., Matsushima, A., Hirano, J., and Inada, Y., *J. Biochem.*, **71**, 897 (1972).
12. Ito, Y., Sisido, M., and Imanishi, Y., *J. Biomed. Mater. Res.*, **23**, 191 (1989).
13. Detwiler, T. C., and Feinman, R. D., *Biochemistry*, **12**, 282 (1973).
14. Phillips, D. R., and Agin, P. P., *Biochem. Biophys. Res. Commun.*, **75**, 940 (1977).
15. Ito, Y., Iguchi, Y., Kashiwagi, T., and Imanishi, Y., *J. Biomed. Mater. Res.*, **25**, 1347 (1991).

11

Acid–Base Effects at Polymer Interfaces

C. J. van Oss

Departments of Microbiology and Chemical Engineering,
State University of New York at Buffalo

1 INTRODUCTION

The interaction between apolar* molecules, surfaces or particles, either *in vacuo* or immersed in apolar liquid media, is solely governed by Lifshitz–van der Waals (LW) forces. Paradoxically, however, the interaction between apolar molecules, surfaces or particles, when immersed in bipolar† media such as water, is mainly governed by polar‡ (Lewis) acid–base (AB) forces, and usually only to a very small extent by Lifshitz–van der Waals forces (and in certain cases to some degree by electrostatic forces, which, however, will not be treated here *in extenso*).

Whilst Lifshitz–van der Waals interactions between similar and dissimilar materials *in vacuo* and between similar materials immersed in a liquid are always attractive, LW interactions between dissimilar materials immersed in a liquid can be attractive *or* repulsive, depending on the LW properties of the materials and of the liquid. On the other hand, polar, electron-acceptor/electron-donor (AB) interactions, even when occurring between *similar* materials immersed in a polar liquid, can be attractive *or* repulsive, depending on the AB properties of the materials and of the liquid. However, AB interactions *in vacuo* cannot be repulsive.

In general, a net repulsion can only occur between two materials 1 and 2, immersed in liquid 3, when an intrinsic property, X, of material 1 is quantitatively larger than the X of the liquid, while that of material 2 is smaller than

---

* Apolar compounds are compounds devoid of any polar (i.e. electron-acceptor and/or electron-donor) or metal property; their cohesive interactions as well as their adhesive interactions (*in vacuo*) are solely governed by Lifshitz–van der Waals forces.

† Bipolar compounds are polar compounds whose cohesion is partly due to electron-acceptor/electron-donor (or Lewis acid–base) interactions.

‡ Polar Interactions are (Lewis) acid–base or electron-acceptor/electron-donor interactions.

Polymer Surfaces and Interfaces II
Edited by W. J. Feast, H. S. Munro and R. W. Richards
© 1993 John Wiley & Sons Ltd

the X of the liquid (or vice versa). It may be held that (in the context of this reasoning) there is only *one* Lifshitz–van der Waals property. This is usually quantitatively expressed as the Hamaker coefficient[1] (generally designated as A_{ii}) of a given material (i). Then a net LW repulsion can only ensue when $A_{11} > A_{33} > A_{22}$ or when $A_{11} < A_{33} < A_{22}$.[2-4] Thus, net LW repulsions, in liquids, are only possible between *different* materials.

On the other hand, in addition to their single LW property (or Hamaker coefficient), polar materials, including polar liquids, have *two* different AB properties in the guise of their electron acceptivity ($\oplus$) and their electron donicity ($\ominus$). When the acceptivity of material 1 is greater than that of the liquid, 3, and its donicity is smaller than that of the liquid (or vice versa), a net AB repulsion ensues. In other words, net AB repulsions will occur when $\oplus_1 > \oplus_3$ and $\ominus_1 < \ominus_3$ or when $\oplus_1 < \oplus_3$ and $\ominus_1 > \ominus_3$.

The mechanism of both types of net repulsions is perhaps better understood by stating that such repulsions only take place when the molecules and/or particles immersed in the liquid are each more strongly attracted to the liquid molecules than to one another. In interactions in polar systems it may well happen that the LW interaction energy (which is always present in apolar as well as in polar systems) and the AB interaction energy have different signs (one may be repulsive and the other attractive, or vice versa). In all cases both modes of interfacial (IF) interaction must be taken into account:

$$\Delta G^{LW} + \Delta G^{AB} = \Delta G^{IF} \qquad (1)$$

Some examples follow of attractive and repulsive LW as well as AB interactions:

(a) *Attractive LW interactions*: miscibility of polymers dissolved in apolar solvents.[3,6]

(b) *Repulsive LW interactions*: phase separation of polymers dissolved in apolar solvents.[3,6]

(c) *Attractive AB interactions*: 'hydrophobic' interactions;[7-9] specific interactions, such as those between antigens and antibodies, enzymes and their substrate, ligands and receptors, and lectins and polysaccharides;[10] 'salting-out'.[11]

(d) *Repulsive AB interactions*: hydration repulsion;[12] hydration forces;[13,14] polymer solubility in water and other polar media;[15] polymer osmotic pressure in aqueous solution;[16] 'steric' stabilization of particle suspensions by neutral polymers in aqueous media;[17-19] phase separation of polymers in aqueous media;[20] depletion flocculation and depletion stabilization;[18,19,21] micelle formation in solutions of nonionic surfactants;[7-22] elution in reversed-phase and hydrophobic interaction liquid chromatography.[11,13,24]

2 THEORY AND METHODS OF MEASUREMENT

2.1 Primary and Secondary Noncovalent Interaction Forces

Only noncovalent forces are considered here; however, for various reasons it is important to define from the outset which of the multitudinous popularly quoted noncovalent forces are genuinely primary and which ones are secondary consequences of one or several primary forces. One important reason for making this distinction is the fact that if one inadvertently were to mistake one of the many secondary phenomena for a real primary force, that secondary force would be counted twice in establishing an energy balance. Of the seventeen forces acting in aqueous media, listed in Table 1, only six can be considered as primary; the other eleven are offshoots of these.

In macroscopic systems, the first three forces, the van der Waals forces, should be grouped together,[28] as they all decay with distance according to the same regimen (at least up to a distance of about 10 nm) and obey the same equations.[26–32] In macroscopic systems, these three van der Waals interactions are alluded to collectively as Lifshitz–van der Waals, or LW, forces.

Electrostatic, or EL, interactions can play an important role when considering, for example, interactions between strong polyelectrolytes in aqueous media. With neutral polymers, EL forces, while rarely completely zero, are nevertheless usually relatively unimportant. For instance, in the case of very hydrophilic neutral polymers, such as dextrans (linear polyglucose molecules), the surface (or $\zeta-$) potential is only about -0.01 to -0.1 mV.[33] In most cases the electrostatic interaction energies (ΔG^{EL}) among polymers with $\zeta \geq |10\text{ mV}|$ may be neglected in comparison with the AB and even the LW interaction energies.

Lewis acid–base (AB) interactions are the predominant forces acting on apolar as well as on polar polymers or particles, immersed in aqueous media. In such media, electron-acceptor/electron-donor interactions mainly manifest themselves as hydrogen-donor/hydrogen-acceptor, or Brønsted acid–base, interactions. In water itself, 70% of its free energy of cohesion is due to AB interactions, and only 30% to LW forces. In 'hydrophobic' interactions among monomeric or polymeric hydrocarbons, in water, the attractive forces for more than 99% are due to AB interactions.[8] Brownian movement forces are rarely of great importance in the interactions between smooth particles or between high molecular weight polymers. In interactions between moieties of lower molecular weight, however, the (repulsive) free energy of Brownian motion ($\Delta G^{BR} \approx +1$ kT) usually needs to be taken into account.

The eleven secondary interaction forces identified in Table 1 are accompanied in that table by a brief summary of their primary origin or origins, and are not further discussed in detail in this chapter. Only the primary noncovalent forces are treated here, with the emphasis on LW and AB interactions, which together

Table 1. Primary and secondary noncovalent interaction forces, acting in aqueous media. (From Ref. 19)

Primary forces	1. London–van der Waals, or dispersion forces	
	2. Debye–van der Waals, or induction forces	Lifshitz–van der Waals, or LW, forces
	3. Keesom–van der Waals , or orientation forces	
	4. Electrostatic or Coulombic forces	El forces
	5. Hydrogen-bonding forces	Lewis acid–base, or AB, forces
	6. Brownian movement forces	BR forces
Secondary interaction forces	7. Hydrophobic interactions	Mainly attractive AB forces
	8. Hydration pressure	Mainly repulsive AB forces, acting at a distance through hydration orientation
	9. Osmotic pressure	In aqueous media, at low concentrations, mainly BR forces; at high concentrations, mainly AB forces
	10. Disjoining pressure	Initially mainly repulsive AB forces;[25] now any or all of the above
	11. Structural forces	Apart from fluctuating forces due to interactions with large molecules of the liquid,[4] mainly repulsive AB forces
	12. Steric interactions	Apart from forces due to chain elasticity,[17] mainly repulsive AB forces
	13. Depletion interactions	Attractive or repulsive AB forces
	14. Entropy-driven interactions	Can only be separately identified after having determined total free energies of interaction at different temperatures
	15. Enthalpy-driven interactions	
	16. Cross-binding interactions	Cross-binding of particles by asymmetric polymer molecules
	17. Specific interactions	Attractive interactions caused by any combination of the first five forces; usually involves sites situated on moieties with a small radius of curvature of 1–10 nm and acting on a small surface area of 0.5–5.0 nm^2 [10]

represent the directly measurable interfacial (IF) forces and which, together with EL forces, are the main interaction forces in condensed media. EL forces have been extensively treated[34–36] (see also Ref. 31 for a brief summary) and will not be elaborated upon here.

2.2 LW Forces—LW Surface and Interfacial Tensions and Hamaker Constants

For all noncovalent cohesive interactions of material i, it may be stated that

$$\Delta G_{ii} \equiv -2\gamma_i \qquad (2)$$

so that also (cf. equation (1))

$$\Delta G_{ii}^{LW} \equiv -2\gamma_i^{LW} \tag{2a}$$

For macroscopic systems,

$$\Delta G_{ii}^{LW} = \frac{-A_{ii}}{12\pi l_0^2} \tag{3}$$

where A_{ii} is the (cohesive) Hamaker constant for material i[1] and l_0 is the minimum equilibrium distance.[37] Combining equations (2) and (3):

$$\gamma_i^{LW} = \frac{A_{ii}}{24\pi l_0^2} \tag{4}$$

Thus, if the easily measured γ_i^{LW} is known for a given material i (see below), then its Hamaker constant, A_{ii}, can be obtained using equation (4), provided that for all materials i the minimum equilibrium distance, l_0, has a constant value. This is indeed the case (at 20°C): $l_0 = 0.157 \pm 0.009$ nm (SD) for 23 materials tested, from liquid helium, through water and various polymers, to mercury,[31] which allows one to express the Hamaker constant as

$$A_{ii} = \gamma_i^{LW} \times 1.86 \times 10^{-21} \text{ J} \tag{5}$$

(where γ_i^{LW} is expressed in mJ/m^2).

The apolar component of the interfacial tension between two condensed-phase materials (i, j), γ_{ij}^{LW}, is

$$\gamma_{ij}^{LW} = (\sqrt{\gamma_i^{LW}} - \sqrt{\gamma_{ij}^{LW}})^2 \tag{6}$$

or

$$\gamma_{ij}^{LW} = \gamma_i^{LW} + \gamma_j^{LW} - 2\sqrt{\gamma_i^{LW}\gamma_j^{LW}} \tag{6a}$$

It is clear from equation (6) that γ_{ij}^{LW} always has a positive value (or is zero).

Equation (6a) is *only* valid for apolar (LW) systems, but Good and Girifalco[38] attempted to expand its use by inserting a factor, Φ, in an attempt to extend the applicability of equation (6a) to polar systems, for which $\Phi \neq 1$:

$$\gamma_{ij} = \gamma_i + \gamma_j - 2\Phi\sqrt{\gamma_i\gamma_j} \tag{7}$$

If, however, equation (7) is solely applied to apolar systems, then

$$\gamma_{ij}^{LW} = \gamma_i^{LW} + \gamma_j^{LW} - 2\Phi^{LW}\sqrt{\gamma_i^{LW}\gamma_j^{LW}} \tag{7a}$$

It can be shown that in apolar experimental systems, $0.98 < \Phi^{LW} < 1.0$,[30,32] which supports the assumption that in macroscopic systems

$$\gamma^{LW} = \gamma^d + \gamma^i + \gamma^o \tag{8}$$

(where the superscripts indicate: d = dispersion, i = induction and o = orientation) is indeed correct.[28,29]

The apolar surface tension component of a solid material (S), γ_S^{LW}, can be determined by means of contact angle (θ) measurements with an apolar liquid (L), for which $\gamma_L = \gamma_L^{LW}$, using the following version of Young's equation:

$$1 + \cos \theta = 2 \sqrt{\frac{\gamma_S^{LW}}{\gamma_L^{LW}}} \tag{9}$$

the main requirement being that $\gamma_L^{LW} > \gamma_S^{LW}$ (see Section 2.5 below). From γ_S^{LW} found in this manner, the Hamaker constant of the solid material can then be obtained, using equation (5).

The apolar free energy of interaction between molecules or particles of a solid or solute (S) immersed in a liquid (L) is expressed as

$$\Delta G_{SLS}^{LW} = -2\gamma_{SL}^{LW} \tag{10}$$

whereas the apolar free energy of interaction between two different kinds of molecules or particles, *i* and *j*, *in vacuo* or in air, is expressed according to Dupré's equation as

$$\Delta G_{ij}^{LW} = \gamma_{ij}^{LW} - \gamma_i^{LW} - \gamma_j^{LW} \tag{11}$$

which, upon combination with equation (6a), becomes

$$\Delta G_{ij}^{LW} = -2\sqrt{\gamma_i^{LW}\gamma_j^{LW}} \tag{12}$$

Thus the apolar interaction energy between two of the same molecules or particles immersed in a liquid (equation (10)), as well as the apolar interaction energy between two different molecules or particles, *in vacuo* or in air (equation (12)), is always attractive, because ΔG_{SLS}^{LW} and ΔG_{ij}^{LW} are always negative (or, in the case of ΔG_{SLS}^{LW}, when $\gamma_S^{LW} = \gamma_L^{LW}$, ΔG_{SLS}^{LW} becomes zero; however, with condensed materials, ΔG_{ij}^{LW} is never zero).

In the case of the interaction between two *different* apolar entities (1 and 2) immersed in an apolar liquid (3), the net free energy of interaction between 1 and 2 *can* be repulsive. The apolar free energy of interaction is described by the condensed-phase version of the Dupré equation (cf. equation (11)):

$$\Delta G_{132}^{LW} = \gamma_{12}^{LW} - \gamma_{13}^{LW} - \gamma_{12}^{LW} \tag{13}$$

which in the light of equation (6a) can be rewritten as

$$\Delta G_{132}^{LW} = 2(\sqrt{\gamma_1^{LW}} - \sqrt{\gamma_3^{LW}})(\sqrt{\gamma_3^{LW}} - \sqrt{\gamma_2^{LW}}) \tag{13a}$$

where ΔG_{132}^{LW} clearly can become positive when $\gamma_1^{LW} > \gamma_3^{LW} > \gamma_2^{LW}$ or when $\gamma_1^{LW} < \gamma_3^{LW} < \gamma_2^{LW}$ or, in terms of Hamaker constants (see equation (4)), a net van der Waals repulsion will occur when $A_{11} > A_{33} > A_{22}$ or when $A_{11} < A_{33} < A_{22}$; see above and also Refs. 1–4 and 31.

2.3 AB Forces—AB Surface and Interfacial Tensions

The importance of making a clear distinction between van der Waals forces and Lewis donor–acceptor interactions at interfaces was first identified by Fowkes.[39] The marked difference between polar, or Lewis acid–base (electron-acceptor/electron-donor or AB) forces, and LW forces, lies in their very different contributions to the interfacial tension, γ_{ij}, and in their different modes of decay as a function of distance between molecules and particles. The crucial difference from LW forces (which, on a macroscopic scale, can be combined into a *single* global property quantitatively expressible as the Hamaker coefficient, which is proportional to γ_i^{LW} in equation (4)) is that AB interaction parameters, even in pure homogeneous materials, comprise two completely different properties which vary independently. These are their electron accepticity ($\oplus$) and their electron donicity ($\ominus$), which as surface tension parameters are designated as the electron-acceptor parameter ($\gamma_i^{\oplus}$) and the electron-donor parameter ($\gamma_i^{\ominus}$). These are not additive[31,40] but relate to the polar, AB, surface tension component as[31,41]

$$\gamma_i^{AB} = 2\sqrt{\gamma_i^{\oplus}\gamma_i^{\ominus}} \tag{14}$$

On the other hand, γ_i^{AB} and γ_i^{LW} *are* additive:

$$\gamma_i^{LW} + \gamma_i^{AB} = \gamma_i \tag{14a}$$

The polar (AB) interfacial tension between materials i and j is then expressed as (see also Small[42])

$$\gamma_{ij}^{AB} = 2(\sqrt{\gamma_i^{\oplus}\gamma_i^{\ominus}} + \sqrt{\gamma_j^{\oplus}\gamma_j^{\ominus}} - \sqrt{\gamma_i^{\oplus}\gamma_j^{\ominus}} - \sqrt{\gamma_i^{\ominus}\gamma_j^{\oplus}}) \tag{15}$$

which also may be written as

$$\gamma_{ij}^{AB} = 2(\gamma_i^{\oplus} - \gamma_j^{\oplus})(\gamma_i^{\ominus} - \gamma_j^{\ominus}) \tag{15a}$$

from which it can be seen that γ_{ij}^{AB} can be *negative*, i.e. when $\gamma_i^{\oplus} > \gamma_j^{\oplus}$ *and* $\gamma_i^{\ominus} < \gamma_j^{\ominus}$ or when $\gamma_i^{\oplus} < \gamma_j^{\oplus}$ *and* $\gamma_i^{\ominus} > \gamma_j^{\ominus}$.

The *total* interfacial tension between materials i and j is then according to

$$\gamma_{ij}^{LW} + \gamma_{ij}^{AB} = \gamma_{ij} \tag{16}$$

where

$$\gamma_{ij} = (\sqrt{\gamma_i^{LW}} - \sqrt{\gamma_j^{LW}})^2 + 2(\sqrt{\gamma_i^{\oplus}\gamma_i^{\ominus}} + \sqrt{\gamma_j^{\oplus}\gamma_j^{\ominus}} - \sqrt{\gamma_i^{\oplus}\gamma_j^{\ominus}} - \sqrt{\gamma_i^{\ominus}\gamma_j^{\oplus}}) \tag{16a}$$

There are many circumstances under which $|\gamma_{ij}^{AB}| > \gamma_{ij}^{LW}$, so that in such cases, when $\gamma_{ij}^{AB} < 0$, the total interfacial tension $\gamma_{ij} < 0$. As we may express the total interfacial free energy of interaction between molecules or particles of material i immersed in a liquid j as

$$\Delta G_{iji}^{IF} = -2\gamma_{ij} \tag{10b}$$

(cf. equation (10)), it becomes evident that when the polar component of the interfacial tension is *predominant and negative*, $\Delta G_{iji} > 0$, which signifies that molecules or particles of material i immersed in liquid j will repel each other, thus giving rise to pronounced solubility of molecules i in liquid j or to marked stability of particles i immersed in liquid j. In both categories the liquid in question is most frequently water, and the solubility of polymer molecules, or the stability of the aqueous suspension of polymer particles in water, is then due to their net mutual hydrogen-bonding (AB) repulsion,[5] even when electrostatically strictly neutral. The equation for these interfacial interactions in liquids (j) is

$$\Delta G_{iji}^{\text{IF}} = -2(\sqrt{\gamma_i^{\text{LW}}} - \sqrt{\gamma_j^{\text{LW}}})^2 - 4(\sqrt{\gamma_i^{\oplus}\gamma_i^{\ominus}} + \sqrt{\gamma_j^{\oplus}\gamma_j^{\ominus}} - \sqrt{\gamma_i^{\oplus}\gamma_j^{\ominus}} - \sqrt{\gamma_i^{\ominus}\gamma_j^{\oplus}})$$

(17)

and the equation for total interfacial interactions between two *different* molecules and/or particles (1 and 2) immersed in liquid (3) is

$$\Delta G_{132}^{\text{IF}} = (\sqrt{\gamma_1^{\text{LW}}} - \sqrt{\gamma_2^{\text{LW}}})^2 - (\sqrt{\gamma_1^{\text{LW}}} - \sqrt{\gamma_3^{\text{LW}}})^2 - (\sqrt{\gamma_2^{\text{LW}}} - \sqrt{\gamma_3^{\text{LW}}})^2$$
$$+ 2[\sqrt{\gamma_3^{\oplus}}(\sqrt{\gamma_1^{\ominus}} + \sqrt{\gamma_2^{\ominus}} - \sqrt{\gamma_3^{\ominus}}) + \sqrt{\gamma_3^{\ominus}}(\sqrt{\gamma_1^{\oplus}} + \sqrt{\gamma_2^{\oplus}} - \sqrt{\gamma_3^{\oplus}})$$
$$- \sqrt{\gamma_1^{\oplus}\gamma_2^{\ominus}} - \sqrt{\gamma_1^{\ominus}\gamma_2^{\oplus}}]$$

(18)

The AB interfacial free energy of interaction between two polar materials (i and j) *in vacuo* must take into account the interaction between the electron acceptor of i and the electron donor of j, *as well as* the interaction between the electron donor of i and the electron acceptor of j (see, for example, Kollman[43]):[31,40,41]

$$\Delta G_{ij}^{\text{AB}} = -2(\sqrt{\gamma_i^{\oplus}\gamma_j^{\ominus}} + \sqrt{\gamma_i^{\ominus}\gamma_j^{\oplus}})$$

(19)

Equation (19) is the crucial AB combining rule which directly gives rise to equations (14)–(18). Using a similar approach, as early as 1953 Small[42] expressed the interaction between two identical polar moieties immersed in a polar liquid with an equation analogous to equation (15).

Comparing

$$\Delta G_{iji}^{\text{AB}} = -4(\sqrt{\gamma_i^{\oplus}} - \sqrt{\gamma_j^{\oplus}})(\sqrt{\gamma_i^{\ominus}} - \sqrt{\gamma_j^{\ominus}})$$

(20)

with equation (13a), it becomes clear that whilst apolar LW interactions can only give rise to a net repulsion when acting between two *different apolar* moieties immersed in an apolar liquid, polar AB interactions can even be repulsive when acting between two *identical polar* molecules or particles immersed in a polar liquid (see Section 1).

The decay as a function of distance, l, of AB energies of interaction between two flat parallel slabs of material is exponential:

$$\Delta G^{\text{AB}}(l) = \Delta G^{\text{AB}}(l_0)\exp\left(\frac{-l}{\lambda}\right)$$

(21)

where l_0 is the minimum equilibrium distance and λ the 'decay length' of the molecules constituting the liquid. For water, λ is of the order of 0.6^{44}–1.0 nm.[45] Thus the regime of decay with distance of AB forces shows a stronger resemblance with the decay of electrostatic interaction (see, for example, Overbeek[34]) than with the decay of LW forces, which is proportional to l^{-2}; see the Hamaker equation (equation (3)) which is also valid for values of $l > l_0$ up to $l \approx 10$ nm (see also Israelachvili[4]).

2.4 Interfacial Tension and Monopolar Surfaces

Using the electron-acceptor/electron-donor approach to interfacial interactions (in particular exemplified by equation (15)), contact angle measurements (see Section 2.5) on surfaces of a variety of polar and, especially, water-soluble polymers and biopolymers revealed that the vast majority of polar polymers have a sizable to very strong electron-donor $(\gamma^{\ominus})$ surface parameter, combined with a (usually) zero or very small electron-acceptor $(\gamma^{\oplus})$ parameter. Such polymer surfaces are best described as 'monopolar surfaces'.[41] Among polymers with totally or predominantly $\gamma^{\ominus}$ monopolar surface properties are (in the approximate order of increasing value of $\gamma^{\ominus}$ [41,46]): polystyrene (here the $\gamma^{\ominus}$ reflects only the π electrons), polymethylmethacrylate, zein (a water-insoluble corn protein), gelatin, cellulose nitrate, cellulose acetate, agarose, serum gamma globulin, serum albumin, polyvinyl alcohol, dextran (a linear polymer of glucose), polyethylene oxide.

Most polymers other than the low-energy polyfluorohydrocarbons and polyhydrocarbons have a value for γ_i^{LW} of the order of 40 mJ/m^2. Thus, with most polymers, a value for $\gamma_i^{\ominus} \geq 28.3$ mJ/m^2 is required for that polymer's total interfacial tension with water (w), γ_{iw}, to be negative (cf. equation (16a)). Thus, when $\gamma_i^{\ominus} > 28.3$ mJ/m^2, $\triangle G_{iwi}^{IF}$ has a positive value, and the polymer is exceedingly soluble in water. This is the case with, for example, (hydrated) serum albumin, polyvinyl alcohol, dextran and polyethylene oxide. For polymers of a sizable molecular weight to be soluble in water, a positive value of ΔG_{iwi}^{IF}, and thus a negative interfacial tension γ_{iw}, is a *conditio sine qua non*. It is therefore not surprising that all water-soluble biopolymers are almost or completely monopolar and have $\gamma_i^{\ominus}$ values equal to or higher than 28.3 mJ/m^2 (especially in the hydrated state). Inspection of equation (16a) reveals that for γ_{iw} to be negative, two things are necessary: (a) a γ_{iw}^{LW} value of $\approx +2.74$ mJ/m^2 must be overcome and, (b) the polar contribution to the cohesion of water $(2\sqrt{\gamma_w^{\oplus}\gamma_w^{\ominus}} = +51$ mJ/m$^2)$ (which is the main cause of 'hydrophobic' attraction) must be surpassed. This is only possible when $\gamma_i^{\oplus} \approx 0$ and $\gamma_i^{\ominus} \geq 28.3$ mJ/m^2 (see equation (16a)).

When on intuitive (but theoretically unfounded[39]) grounds the difference between electron acceptivity and electron donicity is not recognized and a geometric mean combining rule for a polar γ^P component is preferred which is analogous to the one used for γ^{LW}, a γ_{ij} equation of the following shape ensues:

$$\gamma_{ij} = (\sqrt{\gamma_i^{LW}} - \sqrt{\gamma_j^{LW}})^2 + (\sqrt{\gamma_i^P} - \sqrt{\gamma_j^P})^2 \tag{22a}$$

This approach also influences Young's equation (see Section 2.5) which would then take the form of

$$(1 + \cos \theta)\gamma_L = 2(\sqrt{\gamma_S^{LW}\gamma_L^{LW}} + \sqrt{\gamma_S^P\gamma_L^P}) \tag{22b}$$

The approach involving γ^P and a geometric mean (or a harmonic mean) combining rule[47-51] clearly must always give rise to a positive value for γ_{ij}. In aqueous systems, such a positive value of γ_{iw}, which corresponds to a sizable negative value for ΔG_{iwi}^{IF} and thus to a strong attraction, would result in the theoretical prediction of complete *insolubility* (i.e. a solubility smaller than 1 p.p.m. in water of, for example, dextran and polyethylene oxide*), which is, to the contrary, well known to be extremely water soluble (see also Section 4.1). Thus the γ^P approach is not only theoretically too simplistic but it also yields grossly erroneous predictions when confronted with experimental data. Other earlier approaches to the treatment of polar surface forces have been discussed elsewhere.[31,40]

2.5 The Contact Angle as a Force Balance—Young's Equation

In 1805 Young[52] stated in words (but not in actual equations), what may be expressed as

$$(1 + \cos \theta)\gamma_L = W_{SL} \tag{23}$$

where W_{SL} is the work of adhesion between the solid (S) and the liquid (L) of a drop deposited on the solid in the shape of a drop that makes an angle (θ) with the solid at the triple point: air/liquid/solid. $W_{SL} = -\Delta G_{SLS}^{IF}$, combining equations (12) and (19), becomes

$$W_{SL} = 2(\sqrt{\gamma_S^{LW}\gamma_L^{LW}} + \sqrt{\gamma_S^\oplus\gamma_L^\ominus} + \sqrt{\gamma_S^\ominus\gamma_L^\oplus}) \tag{24}$$

so that the complete Young equation may be expressed as[31,41]

$$(1 + \cos \theta)\gamma_L = 2(\sqrt{\gamma_S^{LW}\gamma_L^{LW}} + \sqrt{\gamma_S^\oplus\gamma_L^\ominus} + \sqrt{\gamma_S^\ominus\gamma_L^\oplus}) \tag{25}$$

It is obvious that for any given polar solid there are three unknown entities that are to be determined, i.e. γ_S^{LW}, $\gamma_S^\oplus$ and $\gamma_S^\ominus$. Thus for any solid, contact angles should be measured with at least three different liquids, of which two must be

---

* Using equations (22a) and (22b), and the contact angle data obtained on dextran and polyethylene oxide, with diiodomethane, α-bromonaphthalene, water and glycerol.[15]

polar, using equation (25) three times, in order to be able to solve for the three unknowns. The advancing contact angle should be measured, i.e. the drop should be continuously enlarged (by feeding more liquid through a syringe), until just before the moment of measurement, so that the edges of the drop always touch fresh solid surface, and not one that has been previously wetted by the liquid (as occurs with a retreating drop). However, at the moment of measurement, the drop should no longer actually be advancing. Also, the flat solid surface should be horizontal and *smooth*, i.e. any roughness should have individual roughness site radii smaller than 1 μm.[29]

Upon closer inspection of equation (25), it can be seen that the contact angle can be regarded as the operative parameter of a force balance. The left-hand side of equation (25) represents the *energy of cohesion of the liquid* times a function of the contact angle (θ), which is exactly equal to the *energy of adhesion between the liquid and the solid* given on the right-hand side of equation (25). In other words, the shape of the drop, as manifested by the contact angle (θ), is the result of the equilibrium occurring between the cohesive forces within the liquid (which tend the drop-shape toward a perfect sphere, and a contact angle of 180°) and the adhesive forces between the liquid and the solid (which tend toward total flattening of the liquid drop, or a contact angle of 0°).

Thus, when the surface properties of the liquids are known, the surface properties (γ_S^{LW}, $\gamma_S^{\oplus}$ and $\gamma_S^{\ominus}$) can be found by means of contact angle measurements with a number of different liquids. Table 2 shows the properties of a number of apolar and polar liquids used in contact angle measurement.

2.6 Wicking

Contact angles cannot be reliably obtained on rough surfaces (see above), which makes it difficult to take measurements on particulate or other granular materials. An added difficulty when attempting to determine contact angles on flat layers of particulate materials is that one has to spread them or press them somehow into a flat pellet, which, apart from its surface roughness, has the added drawback of being porous. Thus one is also handicapped by the fact that, during the measurement of a contact angle, the entire drop tends to disappear into the porous pellet. To obviate these difficulties, the contact angle (θ) can be determined by measuring the velocity of capillary rise of various liquids, inside a porous bed or a porous layer composed of the particles on which the surface tension properties are to be determined, by applying Washburn's[59] equation:

$$h^2 = \frac{R + \gamma_L \cos \theta}{2\eta} \tag{26}$$

where h is the height of capillary rise in time t, R is the mean average pore radius of the interstitial pores between particles and η is the viscosity of the

Table 2. List of surface tension components and parameters, and of the viscosities of liquids used for contact angle measurements, at 20 °C

Liquid	γ_L	γ^{LW}	$\gamma^{\oplus}$	$\gamma^{\ominus}$	η^b
			(mJ/m^2)		(poise)
Hexane	18.4	18.4	0	0	0.00326[f]
Octane	21.6	21.6	0	0	0.00542
Decane	23.8	23.8	0	0	0.00907[f]
Dodecane	25.35	25.35	0	0	0.01493
Tetradecane	26.6	26.6	0	0	0.02322[f]
Hexadecane	27.5	27.5	0	0	0.03
cis-Naphthalene decahydride (cis-decalin)	32.2	32.2	0	0	0.0338[f]
α-bromonaphthalene	44.4	44.4	(0)	(0)	0.0489[f]
Diiodomethane (methylene iodide)	50.8	50.8	(0)	(0)	0.028[f]
Ethylene glycol[c]	48.0	29.0	1.92[c]	47.0[c]	0.199
Formamide[d]	58.0	39.0	2.28	39.6	0.0455
Glycerol[d]	64.0	34.0	3.92	5.74	14.90
Water[e]	72.8	21.8	25.5[g]	25.5[g]	0.010

[a] from Jasper.[53]
[b] From Weast,[54] if not otherwise stated.
[c] From Chaudhury[29] and solubility data.[55]
[d] Ref. 56.
[e] Refs. 31 and 41.
[f] From Viswanath and Natakatan.[57]
[g] These are the reference parameters for water; whilst it is not yet possible to establish the ratio $\gamma_L^{\oplus}/\gamma_L^{\ominus}$ with precision for water, the sum of $\gamma_L^{\oplus} + \gamma_L^{\ominus} = 51$ mJ/m² is known.[58] The assumption of $\gamma_L^{\oplus}/\gamma_L^{\ominus} = 1$ facilitates the calculation of γ_i^{AB}, γ_{ij}^{AB}, ΔG_{ij}^{AB}, ΔG_{iji}^{AB} and ΔG_{ikj}^{AB} values. In the results of these calculations, the assumption of $\gamma_L^{\oplus}/\gamma_L^{\ominus} = 1$ *cancels out*.[31,41]

liquid (L). This equation has two unknowns (R and cos θ), but each of these can be solved. First, a number of low-energy liquids should be used, with which $\theta = 0$, so that cos $\theta = 1$, to solve for R.[60,61]* Once R has been found with spreading liquids, cos θ can be obtained by wicking with a number of non-spreading liquids. Table 2 contains a list of the properties of a number of contact angle and wicking liquids. For the purpose of wicking, the viscosities (at 20°C) of these liquids are also listed. It should be noted that glycerol, which is one of the most useful polar liquids for contact angle determinations, cannot be used for wicking because of its extremely high viscosity. Thus, among polar liquids, next to water, formamide and/or ethylene glycol should be used. It should also be noted that liquids that would give rise to contact angles

---

* It could be verified that the value of cos θ indeed remains equal to unity, for all spreading liquids, regardless of their γ_L value. The reason for this was found to lie in the fact that spreading liquids tend to pre-spread to some degree in advance of the actual measurable capillary rise, so that spreading liquids spread over pre-wetted surfaces, i.e. the liquids spread over themselves.[62]

approaching 90°, or higher, cannot be used in wicking, because they will not penetrate the porous aggregate.

Spherical mono-sized particles can be used for wicking by packing them into glass capillary tubes (e.g. of ≈ 1.5 mm interior diameter). Irregularly shaped and sized particles, however, tend to give rise to the formation of skewed liquid fronts when packed into capillary tubes. Such particles can still be used in wicking by resorting to thin layer wicking. This is done by depositing the particles (e.g. from an aqueous or other liquid suspension of about 2 to 5% particles) on to a horizontal glass slide and allowing the liquid to evaporate.[61,62]

3 ATTRACTIVE INTERFACIAL FORCES BETWEEN POLYMERS IN AQUEOUS MEDIA—'HYDROPHOBIC' INTERACTIONS

3.1 Nature of 'Hydrophobic' Attraction Forces

The molecular mechanism of the strong 'hydrophobic' attraction* between apolar macromolecules or particles immersed in water does not yet seem to be established with convincing clarity or unanimity of opinion.[8] According to Israelachvili: '... on the theoretical side the problem is horrendously difficult, and there are no simple theories of the hydrophobic interaction, though a number of promising approaches have been proposed' (Ref. 4, p. 105). Claesson states that 'several years of experimental and theoretical research work concerning repulsive and attractive hydration forces [must be] anticipated before a satisfactory understanding of the origin(s), properties and implications of these forces are obtained' (Ref. 63, p. 82). We showed earlier[30] that 'hydrophobic' interactions consist of both Lifshitz–van der Waals forces *and* hydrogen bonds, where the contribution of the latter interactions is by far predominant. This is made clearer by using equation (16a), which, if material i is strictly apolar and liquid j is water, simplifies to

$$\Delta G_{iji}^{IF} = -2(\sqrt{\gamma_i^{LW}} - \sqrt{\gamma_j^{LW}})^2 - 4\sqrt{\gamma_j^{\oplus}\gamma_j^{\ominus}} \tag{17a}$$

where $-2(\sqrt{\gamma_i^{LW}} - \sqrt{\gamma_j^{LW}})^2$ is the free energy of LW attraction between the apolar and the water molecules and $-4\sqrt{\gamma_i^{\oplus}\gamma_j^{\ominus}}$ is the polar (AB) free energy of cohesion of water, equal to -102 mJ/m². The γ_i^{LW} value of most alkanes is fairly close to ≈ 22 mJ/m²,[53] while γ_j^{LW} for water $= 21.8$ mJ/m². In such cases $\Delta G_{iji}^{LW} = -2(\sqrt{\gamma_i^{LW}} - \sqrt{\gamma_j^{LW}})^2$ is close to zero, so that indeed, for alkanes

---

* As the energy of *attraction to water* of even the lowest-energy polymeric surfaces (e.g. Teflon) still is of the order of 40 mJ/m², the designation 'hydrophobic', which denotes a repulsion for water, is a flagrant misnomer. However, as the term 'hydrophobic' has become encrusted in the scientific literature, and thus is exceedingly familiar to most workers in the field, we are compelled to continue to use it, for the sake of clarity, in quotation marks.

immersed in water, the van der Waals attraction is negligible, leaving only the polar (AB) component of the free energy of cohesion of water $\Delta G_{iji}^{AB} = -4\sqrt{\gamma_j^{\oplus}\gamma_j^{\ominus}} = -102 \text{ mJ/m}^2$.[8] Comparison of the experimental results obtained with Israelachvili's force balance, with alkyl-coated crossed half-cylinders in water,[64] shows that the measured energy of interaction and the energy calculated with the aid of equation (16a) (and taking into account the water contact angle obtained on the alkyl-coated surface[64]) are in fair agreement.[8] In interactions of the type described by equation (17), it is easily seen that the roles of apolar surfaces (*i*) and water molecules (*j*) are mathematically interchangeable[8]:

$$\Delta G_{iji} = \Delta G_{jij} \qquad (17b)$$

Thus, interaction between molecules of one of the two components, in a two-component system, gives rise to an equally strong effective interaction, of the same sign, between the molecules or particles of the other component. One may therefore state verbally that the strong (polar) cohesive attraction between water molecules 'pushes' the other molecules or particles together, with the same energy (see also Tanford[7] and Fowkes[39]).

When instead of identical apolar substances immersed in water, two different substances (1 and 2) are immersed in water (3), equation (18) can be used. In the case of two different apolar substances, the above reasoning, and the results, are entirely similar. It is, however, not universally realized that 'hydrophobic' attractions can also occur between one apolar (1) material and another polar (fairly hydrophilic) compound (2) when immersed in water (3); equation (18) is of course still valid and in most such 'hydrophobic/hydrophilic' cases the sign of ΔG_{132} is still negative.[23,65] Indeed, the very principle of 'hydrophobic' adherence of hydrophilic macromolecules on to largely apolar surfaces, in aqueous media, is the basis for reversed-phase liquid chromatography (RPLC)[23,24] and, after some dehydration by high salt concentrations, for 'hydrophobic' interaction liquid chromatography (HILC).[11,23] The phenomenon is also the major cause of the adsorption of proteins from aqueous solutions on to low-energy polymeric or siliconized surfaces[65] and for the adsorption of PEO on to the surface of polystyrene or other latex particles (cf. Napper[18]).

3.2 Enthalpic and Entropic Contribution to 'Hydrophobic' Interactions

It is widely held that 'hydrophobic' interactions invariably have an entropic origin.[4,7,66,67] However, the understanding that entropic interactions are a major attribute of the 'hydrophobic effect' originates mainly, if not solely, from the increase in entropy accompanying the transfer of *hydrocarbons* from apolar solvents to water.[7]

However, a more general approach to the determination of the interaction energy between apolar or polar molecules (1) immersed in water (3) is to determine the free energy of interaction, ΔG_{131}, between molecules (1) once they are immersed in water (see also Tanford[68]), and to do this not only with hydrocarbons but also with other 'hydrophobic' and even hydrophilic compounds. To determine the enthalpic and entropic contributions of attractive interfacial (i.e. hydrophobic) interactions, as well as of repulsive interfacial (i.e. hydrophilic) interactions, it is indispensable first to ascertain the interfacial free energies, ΔG_{131}, between the apolar (or the polar) molecules or surfaces, as well as the hydration energies, ΔG_{13}, *at different temperatures.*[9] Using alkanes as well as nonalkane apolar (and slightly polar) compounds, it was found that whilst ΔG_{131} for alkanes is indeed predominantly entropic, for other compounds, ΔH_{131} and $T\Delta S_{131}$ contribute about equally (e.g. benzene), and with CCl_4, dibromoethane and heptanoic acid, ΔG_{131} is predominantly *enthalpic*. While also studying the hydration energies (ΔG_{13}) of these and other compounds, it was found that, for example, for alkanes there is a negative entropy of hydration, which is indicative of the fact that alkanes, upon being immersed in water, fit into the water structure in a fairly favourable manner, and thus, upon becoming hydrated, tend to organize the water molecules in their immediate vicinity to a certain degree. Thus, upon bringing two or more such hydrated alkane molecules together (i.e. when they undergo a 'hydrophobic' attraction), they lose part of their (hitherto organized) water of hydration to the bulk water, which then gives rise to the increase in entropy known to accompany 'hydrophobic' interactions among alkanes. However, this is by no means a universal phenomenon among nonalkanes, and to ascertain the degrees of enthalpic and entropic involvement in 'hydrophobic' (and also in other) interactions, the only safe approach is to determine the interaction energies at different temperatures.[9]

The interaction between PEO molecules immersed in water is repulsive (i.e. its ΔG_{131} is *positive*). This (AB) repulsion is completely enthalpic and is in part counterbalanced by an entropic contribution. This is easily understandable in light of the fact that the (mainly enthalpic) energy of hydration of PEO (its ΔG_{13} is negative and has a strongly negative ΔH_{13} component) is accompanied by a fairly strong negative entropy (i.e. $T\Delta S_{13} < 0$), indicative of the orientation of the water of hydration of PEO. Thus, as with alkanes, an attempt to bring two hydrated PEO molecules together causes a loss of some of the organized water of hydration, which then manifests itself as an increase in entropy.

4 REPULSIVE INTERFACIAL FORCES BETWEEN POLYMER MOLECULES OR PARTICLES—HYDRATION FORCES

4.1 Polymer Solubility

The solubility of a solute in a liquid may be expressed as

$$\Delta G_{131} = \ln s \qquad (27)$$

where ΔG_{131} is the free energy of interaction of molecules of material (1) immersed in liquid (3), expressed in units of kT per molecule-pair ($1\,kT = 4.05 \times 10^{-21}$ J at 20°C), and the solubility (s) is given in moles per litre. From equation (27) it is clear that the solubility increases with an increase in the *positive value* of ΔG_{131}. In other words, the more the molecules (e.g. polymer molecules) of a given species *repel* each other when immersed in a liquid, the more they will tend to disperse in that liquid, i.e. the more soluble they will be. At negative values of ΔG_{131} (i.e. when the solute molecules *attract* each other when immersed in a liquid) the solubility decreases fast, as a function of the increasing (negative) value of ΔG_{131} (see Table 3).

Especially for polymer molecules, when the value for ΔG_{131} becomes negative, and is expressed in units of mJ/m^2, the value for ΔG_{131} in kT-units quickly tends to become very negative. This is because the value of ΔG_{131}, expressed in kT, is proportional to the contactable surface area (S_c) that two adjoining polymer molecules have in common, when touching. With most polymers, in the *attractive mode* S_c is at least of the order of 10 nm^2. On the other hand, in the *repulsive mode* with linear polymers, S_c is much smaller, i.e. of the order of 0.21 nm^2 for polyethylene oxide and 0.34 nm^2 for dextran. The much smaller value of S_c in the repulsive mode is due to the very nature of the repulsive interaction (see also Section 4.4.6).

Table 3. Solubility as a function of the free energy of interaction between solute molecules (1) when immersed in a solvent (3) (see equation (27))

ΔG_{131} (kT)	Solubility (s) (mol/l)
4	54.6
3	20.1
2	7.4
1	2.7
0	1.0
-1	0.37
-2	0.14
-3	0.05
-4	0.02
-5	0.007
-6	0.002
-7	0.0009
-8	0.0003
-9	0.0001
-10	0.00005

The conversion of ΔG_{131} from mJ/m^2 to kT units (at $20°C$) is as follows:

$$\Delta G_{131} \text{ (in } mJ/m^2) = (\Delta G_{131} \times S_c)/4.05 \times 10^{-14} \qquad (28)$$

where S_c is expressed in cm^2. It is, however, not always a simple matter to determine the value of S_c, and thus of ΔG_{131}, in units of kT for values for ΔG_{131} (in mJ/m^2) for a number of polymers (e.g. polyisobutylene, polymethylmethacrylate, polystyrene), immersed in various organic solvents.[15] As long as, for a given system, $\Delta G_{131} > 0$, (amorphous) polymers are always readily soluble. When $|-(2 \text{ or } 3 \text{ } mJ/m^2)| < \Delta G_{131} < 0$ there is still usually solubility, but for most systems, when $\Delta G_{131} > |-5 \text{ } mJ/m^2|$, insolubility prevails.[15]

It is a simple matter to correlate the solubility considerations discussed above with the Flory–Huggins[69–71] solubility interaction parameter, χ_{13}.[16]

$$\chi_{13} = \frac{S_c \Delta \gamma_{13}}{kT} = -\ln s \qquad (29)$$

The main difference with earlier approaches, using χ_{13}, is that rather than determining the value for χ_{13} of a given solute/solvent system at the greatest possible dilution of the solute (see, for example, Barton[72]), if one wishes to ascertain the actual value of the solubility, s, of a polymer, it is essential to determine χ_{13} at what is equivalent to the highest possible concentration, i.e. by means of contact angle measurements on the dry material, using equations (16a), (25) and (29).

In completely apolar systems, equation (16c) reduces to

$$\Delta G_{iji}^{IF} = -2(\sqrt{\gamma_i^{LW}} - \sqrt{\gamma_j^{LW}})^2 \qquad (30)$$

which substantiates the validity (for apolar systems) of Hildebrand's solubility parameter (δ).[73] This is because $\delta_i = \alpha\sqrt{\gamma_i^{LW}}$[73] and the solubility in apolar systems is always greatest when the attraction between solute molecules immersed in the solvent is as small as possible. This occurs when γ_1^{LW} or, in Hildebrand's terms, when the difference between δ_i and δ_j is small.

For polymeric surfactant molecules, e.g. polyethylene oxide chains coupled to alkyl groups, in most cases the solubility (s) (equation (27)) may be equated with the critical micelle concentration (CMC).[74] Thus, knowing the surface tension components of nonionic surfactants, their CMC can be calculated. The CMC values calculated in this manner for a wide variety of polymeric nonionic surfactants of different compositions correlates closely with their experimentally obtained CMC values.[22]

4.2 Particle Stability and the DLVO Theory

Whilst the DLVO theory (named after Derjaguin and Landau[75] and Verwey and Overbeek[76]) quite accurately expresses the interaction energies as a function of the distance (l) between particles suspended in apolar liquids, in polar and especially in aqueous liquids there tends to be little correlation between actual particle stability and the stability predicted solely from combined electrostatic and van der Waals interactions. The reason for this is very simple. With 'hydrophobic' particles immersed in water there is, in addition to the two forces treated in the DLVO theory, a very strong 'hydrophobic' attraction (principally of polar, AB, origin; see Section 3), which tends to be one or two orders of magnitude stronger than the van der Waals interaction and is at least of the same order of magnitude as, and frequently stronger than, the electrostatic repulsion. With hydrophilic particles immersed in water there is, in addition to the DLVO forces, a polar (AB) repulsion, which more often than not is not only much stronger than the van der Waals attraction but is also frequently stronger than the electrostatic repulsion.

On the other hand, if energy balances are constructed which comprise AB forces as well as the two DLVO interactions, they can quite accurately predict the degree of particle stability in aqueous media. Thus the observed stability of mildly 'hydrophobic' particles, as a function of ionic strength, correlate well with the energy balances which comprise LW, EL *and* AB forces, while DLVO plots that take only LW and EL forces into account have no correlation with the experimental result.[77] Also, the known stability *in vivo* of blood cell suspensions can only be quantitatively accounted for if one incorporates the (repulsive) AB forces, as well as the LW and EL interactions, into the energy balance.[44,78]

4.3 Hydration Forces

As indicated in equation (21), AB forces decay exponentially as a function of interparticle or intermolecular distance (l) and of the 'decay length' (λ) of the liquid medium. Thus the mode of decay with distance (l) of AB forces is not unlike that of electrostatic forces in liquid media, with the difference that the decay of AB forces is linked to $1/\lambda$ and that of EL forces to the inverse Debye length, κ. As λ is mainly a property of the molecules of the liquid, while $1/\kappa$ depends strongly on the ionic strength, the influence of changes in ionic strength on the decay of AB forces is very slight compared with the influence of ionic strength on the decay of EL forces.

The fact that the decay of AB forces as a function of distance (l) is rather gradual, or in other words that, for example, in water, attractions as well as repulsions due to hydrogen-bonding interactions are still measurable a fair distance away from the surfaces at which these hydrogen-bonding interactions originate, is due to the *orientation of the water molecules of hydration* in the vicinity of (usually) strong hydrogen-acceptor (electron-donor) sites. The repulsion at a distance between strongly hydrophilic surfaces or hydrophilic polymers has been designated as *hydration pressure*, or *hydration forces*.[12,13] The orientation of the water of hydration of very concentrated (human) serum albumin solution could be measured by contact angle determination. At the outer rim of the first layer of hydration, water was found to be about 75% oriented, and at the rim of the second layer the orientation (with the oxygen atom pointing away from the albumin surface) was still about 31%.[79] Thus, whilst 'hydration forces' are not the underlying cause of 'hydration repulsion', they are the means by which monopolar (usually electron-donor) surfaces make their action felt at a distance.

4.4 Phase Separation in Solutions of Two or More Polymers

4.4.1 *Apolar Systems*

When two polymers (1 and 2) are dissolved in an apolar liquid (3) and γ_1^{LW} (or A_{11}) is larger than γ_3^{LW} (or A_{33}) and γ_2^{LW} (or A_{22}) is smaller than γ_3^{LW} (or A_{33}), or vice versa, polymers (1) and (2) dissolved in liquid (3) repel each other, due to a van der Waals repulsion, and phase separation ensues. On the other hand, when the value of γ_3^{LW} (or A_{33}) is either larger or smaller than both γ_1^{LW} and γ_2^{LW} (or than A_{11} and A_{22}), a van der Waals attraction obtains, and complete miscibility is favored (see equation (13a)[3,6]). If one adds a third apolar polymer (4) to a solvent (3) already containing two polymers (1 and 2) which have separated out into two phases, according to the rules mentioned above polymer (4) must *either* join the phase containing (mainly) polymer 1 *or* the phase containing (mainly) polymer 2, but polymer (4) will not initiate a third phase. Thus, depending on the LW properties of polymers *and* solvent, apolar solutions of two or more polymers can give rise to one phase or two (but not more than two) phases.

4.4.2 *Polar Systems Comprising Organic Solvents*

It should be understood that, in addition to polar interactions, apolar interactions always continue to take place in all polar systems. Thus, in polar organic solvents it can readily occur that $\Delta G_{131}^{LW} > 0$ and $\Delta G_{131}^{AB} < 0$ and $\Delta G_{131}^{AB} < 0$, or vice versa. It is only by using the complete equation (18) that one can predict whether miscibility or phase separation will occur.[6] It should be noted that although it is often held that incompatibility among polymer mixtures in solutions (i.e. phase separation) is more the rule than the exception, it can, and indeed does, quite frequently occur that compatibility (i.e. miscibility) prevails. This is easily understood, e.g. in apolar systems, if one realizes that it is at least as likely for the Hamaker constant of the solvent to be either larger or smaller than the Hamaker constants of the dissolved polymers than that the Hamaker constant of the solvent has a value situated exactly in between the values of the Hamaker constants of the polymers. In polar systems this reasoning becomes more complicated, but, nevertheless, the likelihood for phase separation in polar organic media is roughly comparable to the likelihood of compatibility or miscibility. In *very* polar liquids, such as aqueous media, however, it is a rather different matter (see below).

4.4.3   Phase Separation in Aqueous Media–Coacervation

Given the relatively large size of polymer molecules, and thus of their contactable surface areas, S_c, solubility of polymers in water under conditions where $\Delta G_{131} < 0$ is exceedingly improbable. This is because, contrary to apolar solvents and most polar organic solvents, in water there is the additional very strong 'hydrophobic' interfacial attraction, brought about by the exceptionally high polar energy of cohesion of the water molecules. Thus, polymer solubility in aqueous media is in all cases mainly due to the strong monopolar polymer (usually) electron donor $(\gamma_1^{\ominus})$* interaction with the electron acceptor of water $(\gamma_3^{\oplus})$ (cf. the last term $(\sqrt{\gamma_i^{\oplus}\gamma_j^{\ominus}})$ of equation (16a), which is sufficiently dominant to make $\Delta G_{131} > 0$. Then, curiously, the value of S_c becomes somewhat less crucial. Whilst in the attractive mode the S_c value for polymers is always high (when attracting one another, two polymer molecules will strive to stick together over as large an available surface area as possible), S_c values in the repulsive mode tend to be limited to the area defined by two long-chain molecules crossing each other perpendicularly.[16]

Thus, in water, all soluble polymer molecules tend to repel each other, even polymers of one and the same species. Polymer molecules of different species virtually always repel each other with an energy that is somewhat different from the energies with which polymer molecules of the same species repel each other when immersed in water: $\Delta G_{131} \neq \Delta G_{232} \neq \Delta G_{132}$. As different polymers also differ at least somewhat in specific density, phase separation will ensue when aqueous solutions of two (or more) water-soluble polymers are mixed together. Contrary to apolar systems (see above), in aqueous systems *as many different phases can be obtained as there are different polymer species in solution.*[20] Even small, nonpolymeric solutes (electrolytes as well as nonelectrolytes) can form a separate phase, upon admixture to an aqueous polymer solution.[83–85] Aqueous phase separation of polymer–polymer mixtures or polymer–small solute mixtures, in which each constituent migrates mainly to a separate phase, is also called *coacervation.*[85] One must not confound coacervation with *complex-coacervation*, in which two polymers *combine* with each other so that the *complex* is mainly found in one (usually lower) phase, while the other phase becomes almost devoid of either polymer.[83–85]

---

* Practically all water-soluble polymers are monopolar electron *donors*,[15,41,80] with the exception of polyacrylic acid (at low pH).[81–84]

Due to the exponential decay of the orientation of the water molecules of hydration at the surface of water-soluble polymer molecules, the effective monopolar $\gamma_1^\ominus$ parameter decreases in value with distance and thus *ipso facto* also with an increase in dilution. Therefore, a minimum concentration of each water-soluble polymer is required before the repulsion between two different (hydrated) molecules reaches a value (of the order of $\Delta G_{132} \approx 1.0\, kT$) sufficient for phase separation to prevail over Brownian remixing[6,86] (see also Albertsson[20]).

Aqueous phase separation systems, usually comprising the polymers poly-(ethylene oxide) and dextran, are increasingly used for the separation and purification of biopolymers, cells and subcellular particles.[20] This approach is generally alluded to as 'aqueous partition' (see also Walter *et al.*[87]). When aqueous partition is used for the separation of biopolymers, a biopolymer (X) will either have more affinity for the lower (DEX) phase or for the upper (PEO) phase. (Usually the biopolymers to be separated are not applied in high enough concentrations to form a separate phase; see above.) To predict which of the two polymeric phases biopolymer (X) will prefer, equation (18) may be used to calculate both ΔG_{13X} and ΔG_{23X}. Both values are usually positive; the value that is least positive will indicate the compartment to which X preferentially migrates.[86]

REFERENCES

1. Hamaker, H. C., *Physica*, **4**, 1058 (1937).
2. Visser, J., *Adv. Colloid. Interf. Sci.*, **3**, 331 (1972).
3. van Oss, C. J., Omenyi, S. N., and Neumann, A. W., *Colloid Polym. Sci.*, **257**, 737 (1979).
4. Israelachvili, J. N., *Intermolecular and Surface Forces*, Academic Press, London (1985).
5. van Oss, C. J., *J. Dispersion Sci. Technol.*, **11**, 491 (1990).
6. van Oss, C. J., Chaudhury, M. K., and Good, R. J., *Separ. Sci. Technol.*, **24**, 15 (1989).
7. Tanford, C., *The Hydrophobic Effect*, John Wiley, New York (1980).
8. van Oss, C. J., and Good, R. J., *J. Dispersion Sci. Technol.*, **9**, 355 (1988).
9. van Oss, C. J., and Good, R. J., *J. Dispersion Sci. Technol.*, **12**, 273 (1991).
10. van Oss, C. J., *J. Molec. Recogn.*, **3**, 128 (1990).
11. van Oss, C. J., Moore, L. L., Good, R. J., and Chaudhury, M. K., *J. Protein Chem.*, **4**, 245 (1985).
12. Lis, L. J., McAlister, M., Fuller, N., Rand, R. P., and Parsegian, V. A., *Biophys. J.*, **37**, 657 (1982).
13. Parsegian, V. A., Rand, R. P., and Rau, D. C., *Chemica Scripta*, **25**, 28 (1985).
14. van Oss, C. J., in E. Westhof (ed.), *Water and Biological Macromolecules*, Macmillan, Basingstoke, in press.
15. van Oss, C. J., and Good, R. J., *J. Macromol. Soc. Chem.*, **A26**, 1183 (1989).
16. van Oss, C. J., Arnold, K., Good, R. J., Gawrisch, K., and Ohki, S., *J. Macromol. Sci. Chem.*, **A27**, 563 (1990).

17. Ottewill, R. H., in M. J. Schick (ed.), *Nonionic Surfactants*, Marcel Dekker, New York (1967), p. 627.
18. Napper, D. H., *Polymeric Stabilization of Colloidal Dispersions*, Academic Press, London (1983).
19. van Oss, C. J., *J. Dispersion Sci. Technol.*, **12**, 201 (1991).
20. Albertsson, P. Å., *Partition of Cell Particles and Macromolecules*, John Wiley, New York (1986).
21. van Oss, C. J., Arnold, K., and Coakley, W. T., *Cell Biophys.*, **17**, 1 (1990).
22. van Oss, C. J., and Good, R. J., *J. Dispersion Sci. Technol.*, **12**, 95 (1991).
23. van Oss, C. J., Good, R. J., and Chaudhury, M. K., *Separ. Sci. Technol.*, **22**, 1 (1987).
24. van Oss, C. J., *Israel J. Chem.*, **30**, 251 (1990).
25. Derjaguin, B. V., and Obukhov, E., *Acta Physicochim. USSR*, **5**, 1 (1936).
26. Derjaguin B. V., Churaev, N. V., and Muller, W. M., *Surface Forces*, Plenum, New York (1987).
27. Derjaguin, B. V., *Theory of Stability of Colloids and Thin Films*, Plenum, New York (1989).
28. Lifshitz, E. M., *Zh. Eksp. Teor. Fiz.*, **29**, 94 (1955).
29. Chaudhury, M. K., *Short range and long range forces in colloidal and macroscopic systems*, PhD Dissertation, State University of New York at Buffalo, Buffalo (1984).
30. van Oss, C. J., Good, R. J., and Chaudhyry, M. K., *J. Colloid Interf. Sci.*, **111**, 378 (1986).
31. van Oss, C. J., Chaudhury, M. K., and Good, R. J., *Chem. Rev.*, **88**, 927 (1988).
32. Good, R. J., and Chaudhury, M. K., in L. H. Lee (ed.), *Fundamentals of Adhesion*, Plenum, New York (1991), p. 137.
33. van Oss, C. J., Fike, R. M., Good, R. J., and Reinig, J. M., *Analyt. Biochem.*, **60**, 242 (1974).
34. Overbeek, J. Th. G., in H. R. Kruyt (ed.), *Colloid Science*, Vol. I, Elsevier, Amsterdam (1952), p. 245.
35. Hunter, R. J., *Zeta Potential in Colloid Science*, Academic Press, London (1981).
36. Kitahara, A., and Watanabe, A., *Electrical Phenomena at Interfaces*, Marcel Dekker, New York (1984).
37. van Oss, C. J., and Good, R. J., *Colloids and Polymers*, **8**, 373 (1984).
38. Good, R. J., and Girifalco, L. A., *J. Phys. Chem.*, **61**, 904 (1957); **64**, 561 (1960).
39. Fowkes, F. M., *J. Adhesion*, **4**, 155 (1972).
40. van Oss, C. J., Good, R. J., and Chaudhury, M. K., *Langmuir*, **4**, 884 (1988).
41. van Oss, C. J., Chaudhury, M. K., and Good, R. J., *Adv. Colloid Interf. Sci.*, **28**, 35 (1987).
42. Small, P. A., *J. Appl. Chem.*, **3**, 71 (1953).
43. Kollman, P., *J. Am. Chem. Soc.*, **99**, 4875 (1977).
44. van Oss, C. J., in R. Glaser and D. Gingell (eds), *Biophysics of the Cell Surface*, Springer, Berlin, London, New York (1990), p. 131.
45. Marcelja, S., in J. Charvolin, J. F. Joanny and J. Zinn-Justin (eds.), *Liquids at Interfaces*, North-Holland, Amsterdam (1990), p. 102.
46. van Oss, C. J., Good, R. J., and Chaudhury, M. K., *J. Protein Chem.*, **5**, 385 (1986).
47. Owens, D. K., and Wendt, R. C., *J. Appl. Polym. Sci.*, **13**, 1741 (1969).
48. Kaelble, D. H., *J. Adhesion*, **2**, 66 (1970).
49. Hamilton, W. C., *J. Colloid Interf. Sci.*, **47**, 672 (1974).
50. Andrade, J. D., Ma, S. M., King, R. N., and Gregonis, D. E., *J. Colloid Interf. Sci.*, **72**, 488 (1979).
51. Janczuk, B., Chibowski, E., Choma, I., Dawidowicz, A. L., and Bialopiotrowicz, T., *Mater. Chem. Phys.*, **25**, 185 (1990).

52. Young, T., *Phil. Trans. R. Soc. Lond.*, **95**, 65 (1805).
53. Jasper, J. J., *J. Phys. Chem. Ref. Data*, **1**, 841 (1972).
54. Weast, R. C., *Handbook of Chemistry and Physics*, CRC Press, Cleveland (1970), p. F37.
55. Stephen, H., and Stephen, T., *Solubilities of Inorganic and Organic Compounds*, Vol. I, Part 1, Pergamon, Oxford (1963).
56. van Oss, C. J., Good, R. J., and Busscher, H. J., *J. Dispersion Sci. Technol.*, **11**, 75 (1990).
57. Viswanath, D. S., and Natakatan, G., *Data Book on the Viscosity of Liquids*, Hemisphere, New York (1989).
58. Fowkes, F. M., in J. J. Burke, N. L. Reed and V. Weiss (eds.), *Surfaces and Interfacess*, Vol. I, Syracuse University Press (1967), p. 197.
59. Washburn, E. W., *Phys. Rev.*, **17**, 273 (1921).
60. Ku, C. E., Henry, J. D., Siriwardane, R., and Roberts, L., *J. Colloid Interf. Sci.*, **106**, 377 (1985).
61. Giese, R. F., Costanzo, P. M., and van Oss, C. J., *Phys. Chem. Minerals*, **17**, 611 (1991).
62. van Oss, C. J., Giese, R. F., Li, Z., Murphy, K., Chaudhury, M. K., and Good, R. J., *J. Adhes. Sci. Technol*, **6**, 413 (1992).
63. Claesson, P. M., *Forces between surfaces immersed in aqueous solutions*, PhD Dissertation, Royal Institute of Chemistry, Stockholm (1986).
64. Pashley, R. M., McGuiggan, P. M., Ninham, B. W., and Evans, D. F., *Science*, **229**, 1088 (1985).
65. van Oss, C. J., *Biofouling*, (1991).
66. Hiemenz, P. C., *Principles of Colloid and Surface Chemistry*, Marcel Dekker, New York (1986), p. 448.
67. Israelachvili, J. N., Kott, S. J., Gee, M. L., and Witten, T. A., *Langmuir*, **5**, 1111 (1989).
68. Tanford, C., *Proc. Natl Acad. Sci. USA*, **76**, 4175 (1979).
69. Flory, P. J., and Krigbaum, W. R., *J. Chem. Phys.*, **18**, 1086 (1950).
70. Flory, P. J., *Principles of Polymer Chemistry*, Cornell University Press, Ithaca, New York (1950).
71. Huggins, M. L., *J. Chem. Phys.*, **9**, 440 (1941); *Ann. N. Y. Acad. Sci.*, **43**, 1 (1942).
72. Barton, A. F. M., *Handbook of Solubility Parameters and Other Cohesion Parameters*, CRC Press, Boca Raton (1985), pp. 266ff.
73. Hildebrand, J. H., and Scott, R. L., *Solubiliy of Nonelectrolytes*, Reinhold, New York (1950); Dover, New York (1964).
74. Adamson, A. W., *Physical Chemistry of Surfaces*, John Wiley, New York (1982), p. 488.
75. Derjaguin, B. V., and Landau, L. D., *Acta Physicochimica USSR*, **14**, 633 (1941).
76. Verwey, E. J. W., and Overbeek, J. Th. G., *Trans. Faraday Soc.*, **42B**, 117 (1946); *Theory of the Stability of Lyophobic Colloids*, Elsevier, Amsterdam (1948).
77. van Oss, C. J., Giese, R. F., and Costanzo, P. M., *Clay and Clay Minerals*, **38**, 151 (1990).
78. van Oss, C. J., *Cell Biophys.*, **14**, 1 (1989).
79. van Oss, C. J., and Good, R. J., *J. Protein Chem.*, **7**, 179 (1988).
80. Good, R. J., Chaudhury, M. K., and van Oss, C. J., in L. H. Lee (ed.), *Fundamentals of Adhesion*, Plenum, New York (1991), p. 153.
81. Dobry, A., *Bull. Soc. Chim. Belg.*, **57**, 280 (1948).
82. Smith, K. L., Ninslow, A. E., and Petersen, D. E., *Ind. Engng Chem.*, **51**, 1316 (1959).
83. van Oss, C. J., *J. Dispersion Sci. Technol.*, **9**, 561 (1988).
84. van Oss, C. J., *Polymer Preprints*, **32**, 598 (1991).

85. Bungenbeg de Jong, H. G., in H. R. Kruyt (ed.), *Colloid Science*, Vol. II, Elsevier, Amsterdam (1940), pp. 232, 335, 483.
86. van Oss, C. J., Chaudhury, M. K., and Good, R. J., *Separ. Sci. Technol.*, **22**, 1515 (1987).
87. Walter, H., Brooks, D. E., and Fisher, D. (eds.), *Partitioning in Aqueous Two-Phase Systems*, Academic Press, Orlando (1985).

Index